DEDICATION

I dedicate this book to my wife and children who had to put up with a lot of my days and nights away from them, including many weekends, so that I could generate this book. I also dedicate it to my parents who spent a lot of their resources on getting me through my university studies.

T. Khraishi

To the memory of my two colleagues, Kevin Granata and Liviu Librescu, who sacrificed their own lives protecting the students of Virginia Tech on April 17, 2007. Although I will never get to meet them, they have already taught me a lot and for this I humbly dedicate this book to their memory.

M. Al-Haik

EXPERIMENTS IN MATERIALS SCIENCE AND ENGINEERING

Tariq A. Khraishi and Marwan S. Al-Haik

cognella
San Diego, CA

First published in the United States of America in 2011 by University Readers, Inc.

Trademark Notice: Product or corporate names may be trademarks or registered trademarks, and are used only for identification and explanation without intent to infringe.

15 14 13 12 11   1 2 3 4 5

Printed in the United States of America

ISBN: 978-1-60927-868-7

www.cognella.com 800.200.3908

CONTENTS

PREFACE

This is the First Edition of *Experiments in Materials Science and Engineering*. The principal idea behind this textbook arose from the need to consolidate experimental laboratory procedures and materials into a single self-contained and well-explained book. Another reason was to share some uncommon, advanced, and new experiments that today's undergraduate student can be exposed to and benefit from.

The book is divided into 14 chapters, in addition to the preface and dedication parts. The chapters go into sufficient detail on how to perform the different experiments contained within them. This involves a host of illustrations, figures, and equations to better explain the material. Several of those chapters go into common experiments that can be found in a materials science/engineering undergraduate curriculum, or in other engineering curricula. These are things like metallography, macro hardness testing, and tensile testing of materials. This new book, however, discusses experiments not typically seen in undergraduate labs (examples: powder metallurgy, the effect of strain-hardening on material hardness, paper towel tension test, and nano-indentation).

Although many times the experiments or labs call for the availability of certain equipment, machines, and/or apparatuses, several of those are commonly found in materials science and material engineering labs (including mechanical testing and mechanics laboratories) around the nation and around the world. Therefore, it is not difficult to replace your equipment with the ones presented here, since the physical concepts behind such equipment do not change for a given test. Each chapter contains a feature called "*Note to instructors.*" In this feature and within each experiment write-up, several hints and suggestions are provided to replace the exhibited equipment/machines/apparatuses with other suitable replacements.

Lastly, one of the nice characteristics about this book is a feature included in all chapters, called the "*To-Do List.*" This contains a list of points that will guide students to writing technical reports documenting their experiments following each chapter. This is very helpful for instructors, as it saves them the time and effort needed to come up with specific items for students to do following each experiment/lab. It is also helpful for students, as it reminds them of the important points concerning their experiments. To this effect, an entire chapter in this book is dedicated to technical report writing. The chapter on this will set a standard for the students to follow as it teaches them the typical format of a technical report. It is instructional for teachers to spend some time going over this chapter to help the students reach uniformity (not to mention clarity), in their lab report submissions.

We hope this textbook will benefit all those interested in experiments in materials science and materials engineering (including engineering mechanics).

CHAPTER 1
LABORATORY SAFETY

The following are a set of safety rules that are commonly practiced in materials science and engineering labs:

1. Safety glasses/goggles must be worn in all labs when any experiments are running (especially those involving chemical substances and flying/moving objects). Standard or average safety glasses/goggles are adequate for most procedures. However side shields may be necessary for some operations.
2. Rubber gloves should be worn when dealing with any chemical. Also, furnace-rated gloves should be worn when dealing with furnaces.
3. Fume facemasks should be worn when dealing with chemicals.
4. Identify the eye wash areas and lab/building exits closest to you from day one in the lab.
5. Material safety data sheets are available in the lab in case you need to review them.
6. Closed-toe shoes are required at all times in the lab (i.e., no sandals or flip-flops).
7. Long pants or long skirts are required in the lab at all times.
8. No tank tops or tube tops are allowed in the lab. The torso and shoulders must be covered by a shirt, or lab coat. Also very loose sleeves must be rolled above the elbow or restricted so that they cannot become tangled in equipment.
9. Long hair must be tied back while operating machinery.
10. Unsafe behavior including improper use of equipment will not be tolerated.
11. No food or drinks are allowed in the laboratory at any time.
12. The laboratory must be kept clean at all times.
13. Bring your textbook to each lab to reference learning material in case there is a need.

14. Bring yourself to each lab. Attendance is a must by every student and absence will negatively impact your lab grade unless it is excused absence for extenuating circumstance. Moreover, you have to attend on time at the beginning of a lab. Do not come late to labs.
15. Always follow instructions from the lab instructor(s).
16. Failure to abide by these rules will result in loss of laboratory privileges. The instructors may ask anyone to leave at any time for failure to abide by these rules.

Note to instructors: You can augment or edit the above rules in any way that is most suitable for your lab/class situation.

CHAPTER 2
TECHNICAL
REPORT WRITING

GENERAL POINTS

- Assume that the reader has a copy of the lab manual, including figures.
- If you feel it necessary to include part of the manual or its figures, place them in an Appendix (unless their insertion in the text is important for the results).
- Figures in the appendix, however, should not be referenced in the main body of the report. Simply state that more details are given in Appendix A (or B or whatever).

DETAILS

- The report should be detailed to a degree that would allow the reader to rebuild and repeat the experiment, if so desired.
- Reports, in general, are self-contained and require very little outside reading for complete content understanding.

MORE POINTERS

- Be neat and orderly.
- Reports should be typed and checked for spelling and grammatical errors.
- The cover sheet should include:
  - the name of the experiment, the author, date of the experiment
  - other group members, the course title, the class/lab section and group number

- Clearly separate the different sections, using appropriate headings, numbered in accordance with their appearance in the contents page.
- The report shall have a contents page, with each of the sections numbered and listed by page number.
- The main body of the report begins with the Abstract.
- Put the page number (i.e., 1) at the bottom center, halfway up the bottom margin of the Abstract (first page). The Title page and the Table of Contents page are not numbered.
- The appendices each begin on a new page, numbered with a capital letter, a dash, and an Arabic number (i.e., A-1, B-3, etc.).
- All figures, including graphs, shall be labeled with a Figure # and a caption.
- All figures and tables shall be referenced in the body of the report.
- Do not submit glued, taped, or stapled figures on the final report. If need be, photocopy the entire page. Better yet, make a scanned image of the figure and insert into the text. Acknowledge the source of such data.

LANGUAGE POINTERS

- Always avoid using the first and second person (I, we, you, etc.). The third person is appropriate. For example, instead of "We tested steel in uniaxial tension," say "Steel was tested in uniaxial tension."
- Use the past tense at all times unless it's truly awkward. The present tense is acceptable for facts that are generally true.
- For example, it is appropriate to say that "steel is widely used in structural applications because it is both strong and ductile." But when discussing your results, you must use past tense: "the steel tested was found to be both strong and ductile."

REPORT CONTENT

- Abstract
- 1.0 Introduction
- 2.0 Experimental Apparatus and Methods
- 3.0 Results
- 4.0 Discussion
- 5.0 Conclusions
- 6.0 To-Do List
- References
- Appendices

ABSTRACT

- Summarize what was done and what was found in an experiment.
- The abstract should be brief, often only one paragraph, but should highlight all of the relevant information.
- Its heading is centered at the top of the first page

INTRODUCTION

- Any generally accepted equation or fact that is used in the Discussion section to explain your results.
- Summarize the important theoretical assumptions and present the governing equations with some explanation of their origins (a derivation may be appropriate in some cases).
- Theorems which are central to the purpose of the experiment or to the final evaluation of the data.
- Present these theorems in their mathematical forms and define all symbols used.
- All equations should be written on a line of their own with an equation number as is done throughout this manual.
- It is sometimes customary to include a Nomenclature section before the Introduction.

The following is a list of <u>DO NOT INCLUDES</u> in the Introduction:
- Details of the experiment.
- Theory or equation that is not used in the Discussion section.
- The lab manual.
- Paraphrasing of the lab manual is also not acceptable.

EXPERIMENTAL APPARATUS AND METHODS

- Focus on what was actually performed during the experiment.
- Briefly note the important details of the physical set-up of the experiment and the experimental procedure that was followed.
- Describe any particularly important details, such as how and under what conditions the data was collected.
- Especially note any changes or adjustments which were necessary for satisfactory completion of the experiment.
- Generally, the fewer words, the better.
- Do not give the step-by-step details.

RESULTS

- In the form of graphs or tables.
- All graphs and tables should be explained but not discussed.
- Simply state what the graphs and tables are and leave the deductions and relation to theory to the Discussion section.
- All entries must have units. It is best to put the units in the column where the result is identified.
- Tables and graphs must be labeled, numbered, and given a title.

DISCUSSION

- Fully discuss the results in light of the background information and theory discussed in the introduction.
  - Are your results what you expected?
  - Are they consistent with what was discussed the Introduction?
  - Compare the data to all the theory discussed in the Introduction
  - No new theory should be introduced in the Discussion section

- Explain:
  - Any good or bad results
  - Any strengths and weaknesses of the experimental and computational procedures
  - general purpose of the lab and suggestions as to possible applications of the results or further investigations
- Give Sample Calculations for data listed in tables
- These are your individual comments and should reflect any original thoughts or additional results that were obtained
- Do not answer the questions in the lab manuals as part of the Discussion. They should be answered in their own section, after the Conclusions
- Keep in mind that the questions in the lab manual usually do not represent a complete discussion of your data
- Error analyses:
  - Any explanations for the error and suggestions for decreasing the error should be discussed in the Discussion section
  - The detailed calculations of the error analysis need not be discussed in the main body of the report but instead belong in a separate appendix
- If there are very few results or a lot of results it is appropriate to combine the Results and Discussion sections into a section headed Results and Discussion.
- Then you can present your data, state what equation was used to calculate it, and compare it to theory all in one section.

CONCLUSIONS

- A brief summary of the important results and a brief emphasis of the important points of the Discussion.
- No new information or ideas.
- Summarize what you found and what conclusions you drew from it.
- These conclusions should already have been discussed in the Discussion section so only state them here.

REFERENCES

- The literature, to which direct reference was made in the report, shall be placed here.
- Paraphrasing and direct quotes are to be noted by raised numbers or numbers enclosed in brackets in the report. Another common style for referencing includes the use of the first author's (also called the lead author) last name followed by the year of publication of the reference. Such a referencing style can be used instead of the numbering system but they both can't be used together.
- For the numbering system, placement of the literary references shall be in the sequential order of their appearance in the body of the report. For the other system utilizing the last name, references should be listed (i.e., sorted) alphabetically.

Note to instructors: You can augment or edit the above guides or pointers in any way that is most suitable for your lab/class situation. It is also important to determine to the students up front which style of referencing is to be used in their reports.

CHAPTER 3
MEASUREMENTS AND STATISTICS (EXPERIMENT/LAB 1)

Measurements are an important part of engineering. There would be no use in experimenting if there were no way to differentiate the results. Measurements are also important for design and manufacturing. Without accurate measuring instrumentation there is no way to build and reproduce products. In this lab we will discuss several different methods of measuring some specimens of aluminum tubing stock. To measure the stock, rulers, Vernier calipers, dial calipers, and micrometers will be used.

The specimens that are to be measured are shown in the figures below:

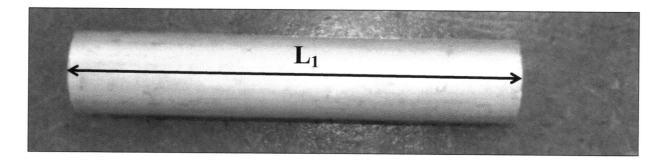

Figure 1. Specimen 1 (tube) Side View

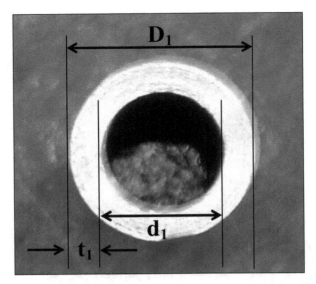

Figure 2. Specimen 1 (tube) Top View

Figure 3. Specimen 2 (tube) Side View

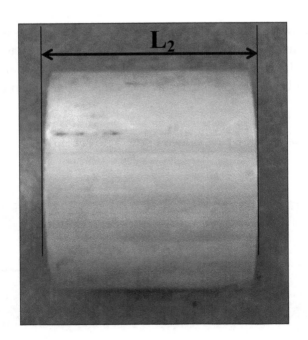

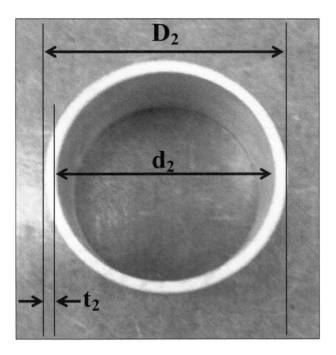

Figure 4. Specimen 2 (tube) Top View

First you should measure and record all dimensions shown on the figures above. These dimensions include length, external diameter, internal diameter, and thickness. Measure these dimensions using a ruler. Once they have been measured using the ruler, then the large dimensions should be measured using Vernier calipers, and the smaller dimensions, such as thickness, should be measured using the micrometer. These values should be recorded in the table below. These values should then be checked using the dial calipers. Once the values are found using each instrument, calculate the volume of material in each specimen. Then calculate the "true percent relative error" using the volume calculated by the ruler measurements as the approximate value and the volume calculated by the micrometer and Vernier calipers as the actual or true value. See the definition of the "true percent relative error" below.

Also measure the mass/weight of each specimen and find from it the density ρ of each specimen. Each student should individually record all of the measurements. Also the "mean" and the "standard deviation" should be calculated (see their definitions below). Once these are calculated check and see if 95% of the measured values fall within mean value $\pm 2S_y$ (S_y being the standard deviation). Then write up a lab report stating the results.

Figure 5. A ruler (in inches)

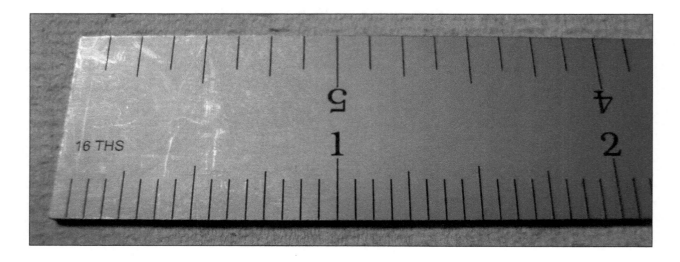

Figure 6. A close-up view of a ruler

- A ruler is an analog device or an analog apparatus (not a digital one), albeit a simple one, that measures the dimensions of an object.
- Figures 5 and 6 show images for a ruler that measures dimensions in inches.
- Each inch, marked on the ruler, is divided into 16 marked divisions on one side of the ruler. On the other side, and for this particular example, each inch is divided into 8 marked divisions. Hence, this particular ruler has two scales on it.
- In this example and using the bottom scale in Figure 6, let's assume that an object's dimension (e.g., length) measures, from the end of the ruler (starting at zero), one whole inch and five and "some" divisions. Let's assume that the "some" here, which represents a fraction of a division, is about half a division (or 0.5 of a division) as seen or estimated by the eye.
- In the above example, the dimension of the object reduces to an addition problem:

---

1 in
+ 5 × 1/16 in (= 0.3125 in)
+ (1/2) × (1/16) in (= 0.03125 in)
1.34375 in

- Note in the above numeric example that the first line of addition (i.e., the 1 in) was a certain quantity or known for sure. The same applies to the second line of addition (i.e., to the 0.3125 in). Therefore, there are four digits after the dot (in the 1.3125 in measurement) that are significant digits (i.e., their inclusion is both important and meaningful).

- The third line, however, was an estimate albeit a good estimate (done with our own analog measuring tool: the eye). Including the last line in the addition process produces a fifth digit after the dot in 1.34375 and actually alters the last three digits in the original 1.3125 number.

- In other words, the last four digits (from the right) in the 1.34375 measurement are not certain. From the point of view of writing decimal numbers, one would instead write this measurement as 1.34 inches (instead of 1.34375 in). The second digit in 1.34 after the dot (i.e., the "4") is also termed a significant digit although its exact quantification is uncertain. Although its exact quantification will depend on who did the eyeballing, most people will estimate the fraction to be 0.5 of one sixteenth of an inch leading to the "4" in the second digit after the dot for the 1.34 dimension measurement here.

- Therefore in this problem there are three significant digits in the 1.34 in measurement and specifically two significant digits after the dot. Trying to figure out what digit comes after the "4" here is extreme speculation and has no significance using this measuring tool. In fact, adding any other digit after the "4" would cause that digit to be called an "insignificant" digit as this would be tantamount to an estimation process over an already estimated process involving the second digit after the dot.

- In conclusion thus, the number of significant digits after the dot are governed by the last digit that can be reasonably estimated.

- Although the above example dealt with inches (which are part of the US Customary Units), it is equally applicable to SI units (e.g., meter/m, centimeter/cm or millimeter/mm).

- Figure 7 shows another quick example illustrating "significant digits." Assume the shown scale in the figure is in cm. If one wants to determine where the arrow in the figure points to, it can be reasonably estimated that the arrow points to 2.82 or 2.83 (note that each cm is divided into ten divisions with each division represent 0.1 cm). Either of these answers would be appropriate and both of them contain two significant digits after the dot. On the other hand, one cannot say that the arrow points to 2.825, for example. Since the "2" after the dot is uncertain, i.e., has some error associated with it, the "5" after the "2" has even more error or uncertainty associated with it and therefore it is not a significant digit and should not be written in formal technical communications.

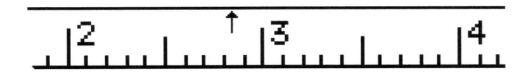

Figure 7. Dimension measurement with significant digits

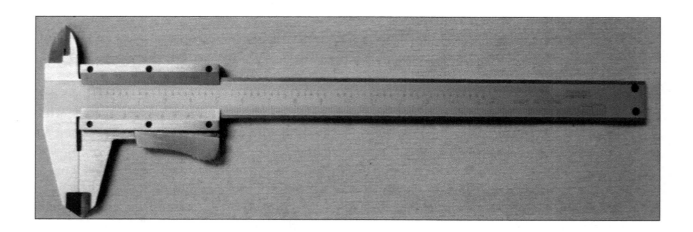

Figure 8. Vernier caliper

Figure 9. A close-up view of a Vernier caliper

- The Vernier caliper is a tool or an apparatus/device that is used to measure dimensions. It can be used to measure external diameter, internal diameters, lengths, and depths. It comes in analog or digital forms, although the analog version is the most common since it is less costly. Its accuracy is typically 0.02 mm or 0.001 in (i.e., 0.001"). Some Vernier calipers can have an even greater accuracy (e.g., 0.0005"). It is, therefore, considered a precise measuring instrument.
- Close the jaws *lightly* on the object to be measured.
- If you are measuring an object with a round cross-section, make sure that the axis of the object is perpendicular to the caliper jaws. This is necessary to ensure that you are measuring the full diameter and not merely a chord.
- Typical Vernier calipers have two scales engraved or printed on them. One is in inches (US Customary units) and one is in centimeters (metric units).
- There are also a fixed scale and a sliding scale for each of these unit systems.
- To illustrate the use of a Vernier caliper in dimension measurement, consider Figure 9. In this figure, we are going to focus on the bottom scale (metric in cm and fractions of a cm, i.e., mm).
- In Figure 9, the fixed scale is read as a ruler is read. In this case, the opening of the caliper jaws around the object's dimension points to a measurement of 1.3 cm (1 cm + 3 × 0.1cm) since there is one whole cm and 3 unmarked divisions (each division representing one tenth of a cm) on the fixed scale. The arrow for the 1.3 cm reading is indicated or pointed out by the 0 on the sliding scale.
- The sliding scale further subdivides the unmarked divisions on the fixed scale. This is where the superiority of the Vernier caliper lies over a ruler. This is a feature that is not present in regular rulers. In regular rulers one would have to estimate a third digit (a second digit after the dot) at this juncture to finish the measurement. With the Vernier caliper, the measurement is still not over for this example.
- In the example caliper shown above, the sliding scale subdivides each 0.1 cm on the fixed scale to ten marked divisions (each worth one tenth of 0.1 cm or 0.01 cm) on the sliding scale. Furthermore, each marked number on the sliding scale is subdivided into five unmarked divisions. In other words, each unmarked division on the sliding scale is worth (1/5)×0.01 cm = 0.002 cm = 0.02 mm.
- The next step in the measurement is to look at the tick marks on the sliding scale and determine (by the eye) where one of them exactly lines up with a tick mark on the fixed scale.
- In the above example, the third tick after the number 4 on the sliding scale seems to be the one of interest.
- The number 4 on the sliding scale is worth 4 × 0.01 cm (= 0.04 cm) towards the measurement. The three ticks after the 4 on the sliding scale are worth 3 × 0.002 cm (= 0.006 cm) to the measurement.
- Hence our measurement reduces down to an addition problem:

1 cm

+ 0.3 cm

+ 0.04 cm

+ 0.006 cm

———————

1.346 cm

- Notice that in this problem, and using this particular Vernier caliper, there can only exist three significant digits after the dot. The third or last digit is obtained by reasonable estimation of where a tick on the sliding scale meets a tick on the fixed scale. Since this is an eyeballing technique that involves error, albeit a reasonable technique, no further estimation of digits can be meaningful beyond this digit. Hence, this problem is limited to three significant digits after the dot.

- On those rare occasions when the reading just happens to be a "nice" number like 4 cm, don't forget to include the zero decimal places showing the precision of the measurement and the reading error. So not 4 cm, but rather 4.000 cm.
- The same method is used for English units.

Using a Micrometer:

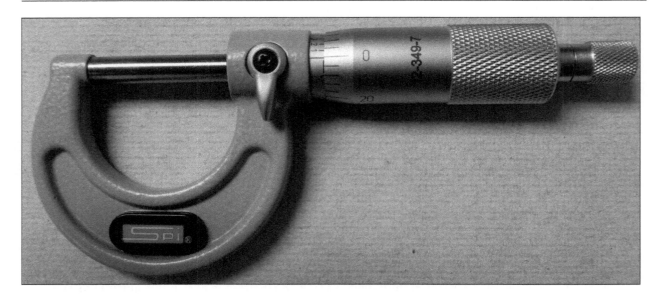

Figure 10. A micrometer

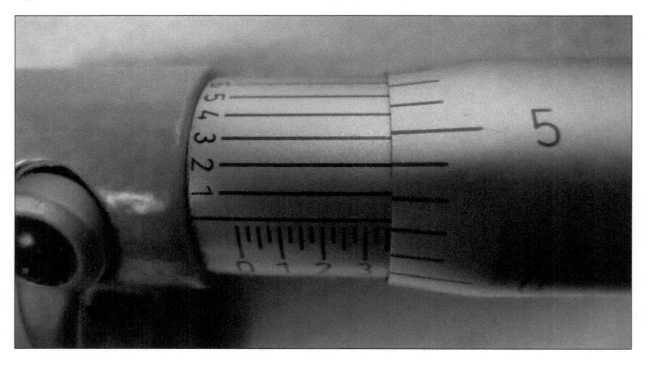

Figure 11. A close-up view of a micrometer

- The micrometer is another precise measuring instrument, similar to the Vernier caliper, which can be used to measure dimensions (typically external diameters and thicknesses). It is typical for its accuracy to be 0.0001 in. However, the accuracy can vary with the exact micrometer and normally goes down with an increase in the range of measurement of the micrometer. Micrometers can be analog or digital although the analog variety is more common, due to it being less expensive.
- Use the ratchet knob (at the far right in Figure 10 above) to close the jaws *lightly* on the object to be measured. It is not a C-clamp! When the ratchet clicks, the jaws are closed sufficiently.
- The micrometer utilizes measurement concepts similar to the Vernier caliper.
- To illustrate how to read a measurement off a micrometer, consider Figure 11. It is a micrometer that reads in inches.
- In Figure 11, the jaws are closed on the dimension to be measured. In this figure, there are three scales on the micrometer. One of the scales is along the barrel of the micrometer and is fixed (shown horizontal in the figure). There is a rotational scale that rotates with the knob rotation (on the right side of Figure 11). Lastly, there is a third scale that is fixed and goes around the barrel (above the horizontal fixed scale mentioned first above).
- In this example, we first read the horizontal fixed scale (which is divided into numbers each representing 0.1 in. Each of these numbers is divided into four divisions each representing 0.25 × 0.1 in = 0.025 in). It is pointing (at the edge of the rotating scale) to 0.3 in + 2 × 0.025 in = 0.350 in.
- The rotating scale is divided into 25 divisions. The whole range (i.e., the whole 25 divisions together) represents a subdivision of each of the small divisions on the fixed horizontal scale (i.e., represent 0.025 in). Therefore, each division on the rotating scale represents 0.001 in. Since two full divisions on the rotating scale have traveled below the horizontal line on the fixed horizontal scale, these two divisions add 2 × 0.001 in (= 0.002 in) to the measurement. The measurement now stands at 0.352 in.
- The last significant digit in the measurement comes from looking at the ticks on the second scale and the ticks on the third scale and seeing which ones line up. There are ten divisions on the third scale and these subdivide each of the divisions on the rotating scale. Hence, each division on the third scale is worth 0.1 × 0.001 in (= 0.0001 in).
- Since in the above example the second tick on the third scale lines up with a tick on the rotating scale, we need to add the following to the measurement: 2 × 0.0001 in (= 0.0002 in).
- Therefore, the measurement in this example boils down to the following addition:

0.3 in
+ 2 × 0.025 in (= 0.050 in)
+ 2 × 0.001 in (= 0.002 in)
+ 2 × 0.0001 in (= 0.0002 in)

———————————————

0.3522 in

- Notice that the number of significant digits after the dot in this case is four.
- In the example above the units are in inches. However, the procedures are exactly the same for SI units.
- The micrometer may not be calibrated to read exactly zero when the jaws are completely closed. Compensate for this by closing the jaws with the ratchet knob until it clicks. Then read the micrometer and subtract this offset from all measurements taken. (The offset can be positive or negative.)
- On those rare occasions when the reading just happens to be a "nice" number like 4 in, don't forget to include the zero decimal places showing the precision of the measurement and the reading error. So not 4 in, but rather 4.0000 in.

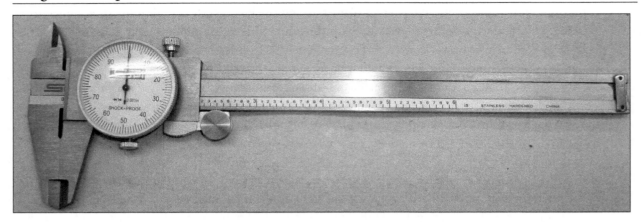

Figure 12. A dial caliper

Figure 13. A close-up view of a dial caliper

The usage of a dial caliper is similar to the Vernier caliper except instead of the Vernier scale, there is a dial scale that gives the smallest readings of the instrument. In the figures here, each 0.1 in on the linear fixed scale (the smallest division on this scale) is divided into 100 divisions on the dial. Therefore, each division on the dial is equal to 0.1 in / 100 = 0.001 in, which is the accuracy of this particular dial caliper.

	Ruler (units)	Vernier Calipers (units)	Micrometer (units)
L_1			Depends on the micrometer range
D_1			
d_1			N/A
t_1			
L_2			Depends on the micrometer range
D_2			Depends on the micrometer range
d_2			N/A
t_2			

Mass (m_1) of specimen 1, units	
Mass (m_2) of specimen 2, units	

Student Name: **Date:**

STATISTICS

Here is some statistical background that will be needed to complete labs. The reason that statistics is needed is that for a measured quantity such as the diameter of a shaft, there will be some variability in the measurements of this quantity depending on several factors (e.g., who made the measurement, the amount of pressure applied on the micrometer/caliper, etc.). A quantity that has a scatter in its determination is called a statistical variable. This quantity can be denoted by y_i, where the subscript i, which ranges from 1 to n, indicates the measurement number, and n is the total number of data points or measurements.

A measure of the central tendency of a sample of data points is the "arithmetic mean" (or "average") y_{avg} defined as:

$$y_{avg} = \frac{\sum_{i=1}^{n} y_i}{n} \qquad (1)$$

A measure of the degree of the spread of the data points around the mean is given by the "standard deviation" S_y defined as:

$$S_y = \sqrt{\frac{\sum_{i=1}^{n} (y_i - y_{avg})}{n-1}}, \text{ where } n \geq 2 \qquad (2)$$

and

$$S_t = \sum_{i=1}^{n} (y_i - y_{avg})^2 \qquad (3)$$

is called the total sum of the squares of the residuals between the data points and the mean.

Notice that

$$\sum_{i=1}^{n} (y_i - y_{avg}) \qquad (4)$$

without squaring is not a helpful measure of spread because it always evaluates to zero.

The reason that the standard deviation is divided by n-1 (Equation 2) and not n is that in order to have a meaningful spread of data, n must be greater or equal to 2, otherwise a meaningless result of infinity is obtained. In other words, you cannot talk about a spread unless you have more than one data point. Another way to think about it is that the y_i's are related to the y_{avg}, and hence our degrees of freedom in S_t are really n-1 not n.

The "variance" is defined as the square of the standard deviation:

$$(S_y)^2 = \frac{S_t}{n-1} \qquad (5)$$

Next the "coefficient of variation" (c.v.) is defined as:

$$c.v. = \frac{S_y}{y_{avg}} \times 100\% \qquad (6)$$

which represents a normalized measure of the spread. This c.v. is similar in spirit to ε_t (the true percent relative error) in that it is a measure of the ratio of the error to the estimate of true value (i.e., y_{avg}). ε_t in general is defined as:

$$\varepsilon_t = \frac{\text{approximate value} - \text{true value}}{\text{true value}} \times 100 \tag{7}$$

Alternatively, ε_t can be defined as:

$$\varepsilon_t = \frac{\text{measured value} - \text{actual value}}{\text{actual value}} \times 100 \tag{8}$$

ε_t is used to quantify the amount of error in percentage that a measurement or approximation yields *knowing* what the true or actual value is. *If* the true or actual value is not precisely known, it can be replaced in this equation by the best-known value for the quantity at hand.

If a statistical quantity is determined in a completely random fashion using a large sample of data (i.e., $n > 30$ but preferably for $n > 40$), then for a quantity that is "normally distributed," i.e., one whose data follows a "normal distribution," there is a "confidence" level that 68% of the data lies between $y_{avg} - S_y$ and $y_{avg} + S_y$ and that 95% of the data lies between $y_{avg} - 2S_y$ and $y_{avg} + 2S_y$.

There is a unique graph shape for normally distributed data that manifests in it the confidence levels discussed above. If one plots what is called a "histogram" for the data, the plot for normally distributed data looks like a symmetric bell-shaped curve (for a sufficient or large enough number of data points). The way a histogram plot is constructed is the following. Divide the data into intervals of equal size. Count the frequency of occurrence of the data within each interval. For example, assume that several people each made several measurements for length L_1, say using a ruler. Assume you have sufficient number of measurements (e.g., 40 or larger). You can then construct a table of data and see how many data points fall into the ranges: 50.53 mm–50.54 mm, 50.54 mm–50.55 mm, 50.55 mm–50.56 mm, 50.56 mm–50.57 mm, 50.57 mm–50.58 mm, 50.58 mm–50.59 mm, and 50.59 mm–50.60 mm. Let's assume that for the ranges exhibited, the number of data points falling in such ranges, respectively, is 5, 10, 18, 25, 19, 9, and 4. If the frequency is plotted on the ordinate and the ranges plotted on the abscissa, one obtains the histogram in Figure 14. For normally distributed data, the center of this 2D curve on the abscissa should be where the data mean lies. Also, for normally distributed data, within a standard deviation on either side of the center should lie 68% of the data, and within two standard deviations should lie 95% of the data. As the number of points keeps growing, the size of each range will go down to one point in the limit.

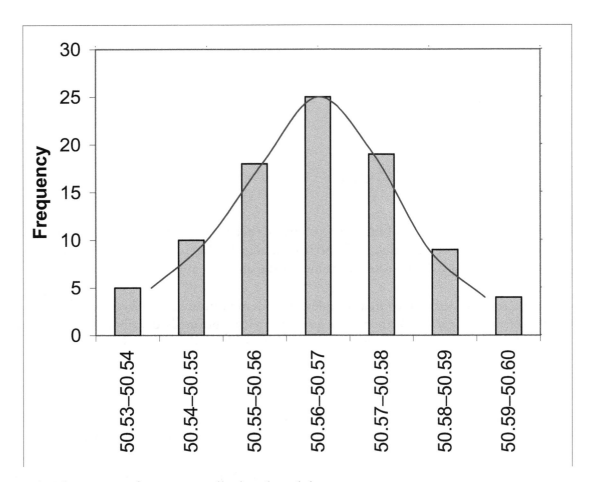

Figure 14. A histogram of some normally distributed data.

The reason for the above limited n statistics is to infer the statistical properties of a population (i.e., the true mean μ and the true standard deviation σ) based on a limited sample of data having a mean y_{avg} and a standard deviation of S_y. This is called the "statistical inference." As expected as the number of data points n increases (in the sample), y_{avg} approaches μ and S_y approaches σ.

TO DO LIST

1. Which one of the measuring instruments allowed for the most significant digits? Explain why.
2. List three different sources of error that could have caused varying results in this experiment. Describe each source.
3. Pick a set of measurements (one that includes a micrometer reading). Calculate average and standard deviation for these measurements.
4. Which measuring instrument yielded the highest standard deviation? Explain why.
5. How are the data points distributed? Does it follow a pattern?
6. For the following set of data points, representing some measurement with error, plot a histogram and include it in your report.

3.567
3.719
3.779
3.786
3.799
3.803
3.810
3.841
3.850
3.854
3.867
3.882
3.889
3.889
3.899
3.901
3.903
3.906
3.922
3.924
3.933
3.938
3.966
3.996
4.014
4.018
4.025
4.030
4.054
4.172

7. For the following two pictures of a Vernier caliper, determine the reading for each based on its jaw opening. Determine using an addition problem and using the US Customary Units scale on the caliper.

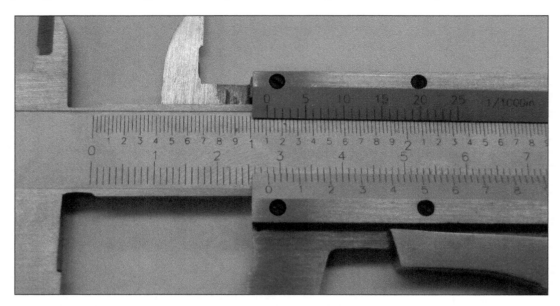

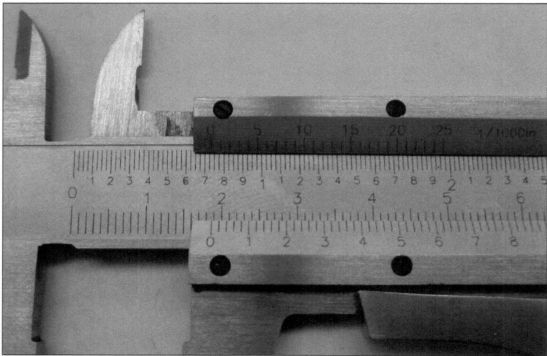

Note to instructors: You can augment or edit the above procedures or To-Do List in any way that is most suitable for your lab/class situation. Note that the exact instruments used will vary but the principles taught here remain the same. To get a large set of data that is amenable to statistics, you can ask each student to do several measurements and record them. Afterwards, data collected by *all* students can be gathered and put together for statistical analysis. Note also that not all data follow a "normal distribution."

CHAPTER 4
CRYSTALLINE STRUCTURE (EXPERIMENT/LAB 2)

The objective of this lab is for students to become familiar with crystal structures of various solids through constructing these structures themselves using ping-pong balls as atoms.

INTRODUCTION

Solid materials may be classified according to the regularity with which atoms or ions are arranged with respect to one another. A crystalline material is one in which the atoms are situated in a repeating or periodic array over large atomic distances; that is, long-range order exists, such that upon solidification, the atoms will position themselves in a repetitive three-dimensional pattern, in which each atom is bonded to its nearest-neighbor atoms. All metals, many ceramic materials, and certain polymers form crystalline structures under normal solidification conditions.

Some of the properties of crystalline solids depend on the crystal structure of the material, the manner in which atoms, ions, or molecules are spatially arranged. There is an extremely large number of different crystal structures all having long-range atomic order; these vary from relatively simple structures for metals to exceedingly complex ones, as displayed by some of the ceramic and polymeric materials. The present discussion deals with several common metallic crystal structures.

When describing crystalline structures, atoms (or ions) are thought of as being solid spheres having well-defined diameters. This is termed the atomic hard sphere model in which spheres representing the nearest-neighbor atoms touch one another.

We will look at three crystal structures: body-centered cubic (BCC) crystal structure; face-centered cubic (FCC) crystal structure; and hexagonal close-packed (HCP) crystal structure. These represent most metals and are relatively simple crystal structures. We will use ping-pong balls to represent the atoms and hot glue to bond the atoms together in a crystal structure.

Unit Cell

To characterize crystal structures, we choose a small group of atoms that form a repetitive pattern. This is called the unit cell. A unit cell represents the building block of a 3D crystalline array or of a crystal (a "grain" is another term for "crystal"). In choosing the unit cell for a particular structure, all defining features must be included along with a high level of symmetry. Using the unit cell, we can generate the entire structure by translating the cell itself along all of its edges repeatedly.

Atomic Packing Factor

The atomic packing factor is the sum of the sphere volumes of all atoms within a unit cell divided by the unit cell volume.

$$APF = \frac{\text{volume of atoms in a unit cell}}{\text{total unit cell volume}} \qquad (1)$$

Equipment and Materials

- Hot glue gun (also called Hot Melt Glue Gun)
- Ping-Pong Balls
- Gloves

Procedure

- Plug in the hot glue gun at your workstation a least 3 minutes before use to let the heating element rise to operating temperature.
- Count out the number of ping-pong balls required to complete the model.
- Glue the ping-pong balls together one at a time, allowing enough time in between for the glue to set (refer to the figures for example models).
- Take a picture/photo of completed model and include in your lab report.

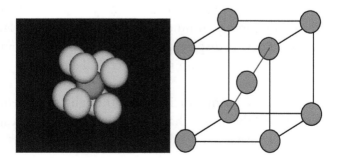

Figure 1. Two BCC models (one with atoms touching and one with them separated).

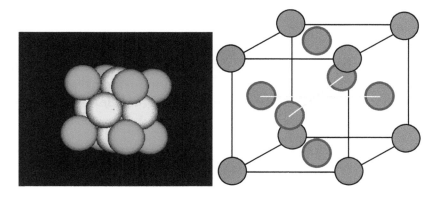

Figure 2. Two FCC models (one with atoms touching and one with them separated).

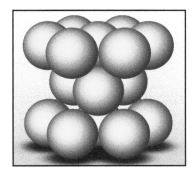

Figure 3. HCP model (with atoms touching).

Figure 4. Rhombohedral (Trigonal) model (with atoms touching).

TO DO LIST
- Using 3D CAD software, draw in 3D the unit cells for BCC, FCC, and HCP structures.
- Calculate the atomic packing factor for the BCC, FCC, and HCP structures (show work).

- Having the information from questions 1 and 2, what other useful property can we then obtain for a given metal with these crystal structures? Explain.

Note to Instructors: You can augment or edit the above procedure or To-Do List in any way that is most suitable for your lab/class situation. Other spheres besides ping-pong balls maybe used in this lab. The issue is to find a glue that would adhere nicely to such balls. Other ideas is for students to take close-up photos of salt and sugar crystals and include in their reports.

CHAPTER 5
METALLOGRAPHY
(EXPERIMENT/LAB 3)

The purpose of this lab is to examine the microstructure of brass or aluminum 6061-T65 through the metallography process.

INTRODUCTION

Metallography is the science of interpreting and reporting the microstructure of engineering materials. Sectioning, mounting, grinding, and polishing are the sample preparation steps prior to the microscopic evaluation. Metallography reveals information about:

- The internal structure of the metal; grain density and size;
- Existence of different phases;
- Existence of impurities; and
- Existence of defects, damages, or change due to loading, corrosion, and/or elevated temperature.

The size of grain has a great effect on the mechanical properties of metals. During the heat treatment, heating time, temperature, and alloying elements all affect grain growth. In all metals a small grain is generally preferable to a large grain. Small grain metals have greater tensile strength and hardness, and also distort less during quenching. As seen under a microscope, the grain size is determined by the count of grains per square inch under 100X magnification. The metallurgical microscope is used to determine the structure of metals. Metal characteristics, grain size, and heat treatment effects on content may be determined by microscopic evaluation.

Sample

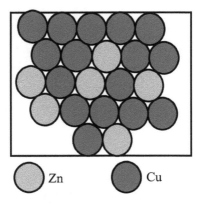

Figure 1. Brass as substitutional solid solution.

One of the material samples used here is cartridge brass. Cartridge brass is an alloy of copper and zinc processed mostly through casting. This material has an appealing yellow color and is used in many structural applications. Both copper and zinc are generally ductile with copper being highly conductive compared to zinc. Copper has a FCC structure and zinc has an HCP structure. They both have comparable atomic radii and electronegativity. Cartridge brass is a solid solution of 70% wt copper and 30% wt zinc. It is a single phase metal (a terminal phase on the phase diagram). This material is stronger than both pure copper and zinc. But it is slightly more conductive than zinc. Cartridge brass is called a substitutional solid solution because some copper atoms are substituted with zinc atoms (Figure 1). Zinc atoms are slightly larger than copper (R_{Zn} = 0.133 nm, R_{Cu} = 0.128 nm). So their intermixing causes minor distortion of the lattice.

Apparatus and Equipment

- Particle mask, goggles/eyewear, and gloves
- Metal cut-off saw or hacksaw
- Specimen mounting mold
- Disk grinder/ polisher
- Pressurized air nozzle
- Light metallurgical microscope with digital camera

Materials

- 6061 Aluminum samples / brass samples
- Distilled water
- Acetone
- Chemical etchant
- Alumina/diamond paste/spreader (extender)
- Silicon carbide abrasive papers (120–1000 grit)
- Polishing cloth

The main reason for this stage is to accommodate comfortable handling of a small sample.

Proper preparation of the sample includes the following stages:

1. Cutting a sample from a bulk piece. The metal sample typically ranges from approximately the size of a pencil eraser to the size of a dime. Larger pieces are not generally any more useful.
2. Mounting the sample in an epoxy resin to accommodate safe and proper handling.
3. Grinding (multiple stages), to remove damaged layers.
4. Polishing (multiple stages), to remove damage due to grinding
5. Chemical etching, to expose the grains and micro features.

PROCEDURES

Stage 1: Cutting

The specimen to be studied must be cut using the metal cut-off saw or a hacksaw. Cutting must be done with coolant flowing on the cutting wheel, to avoid excessive specimen heating. Overheating may cause tempering, thus modifying the processed sample.

Hacksaw:

1. Wear the goggles, mask, and the latex gloves.
2. Set the material sample in the table-mounted vise.
3. Make sure vise is securely clamping the sample.
4. Use hacksaw to make a notch for desired cut location.
5. Use hacksaw to cut sample at the notch (it's easier to cut at an angle).
6. Deburr sample with metal file.
7. Remove original sample from vise and take the cut piece to the epoxy molding station.

Metal cut-off saw:

1. Wear the goggles and the latex gloves (the saw compartment itself protects against flying dust).
2. Set the material sample in the cutting machine.
3. Close the plastic cover.
4. Activate the cutting disk by pressing the cutter button (Green).
5. Pump in the cooling fluid by pressing the coolant button (Blue).
6. Start the cutting process by using the handle attached to the machine (to your right).
7. Press the stop button (Red) and **wait till the cutting disk stop rotating**. Remove the plastic cover and take out your sample. **Sample might be hot so be careful when trying to take it out.**
8. The material sample should be held by a mold, so the next step would be molding.
9. Turn off the saw/cooler pump after you are done.

Stage 2: Mounting

1. Wear goggles and latex gloves.
2. Get a plastic cup, two molding cups, and a stirring stick. Place them all on a piece of paper (e.g., paper towel).
3. Place the material sample **in the center** of the molding cup at its bottom.
4. Mix epoxy resin: two parts Fast-Acrylic powder to one part liquid in the plastic cup.

5. Stir the mixture well for about 15 seconds using a disposable stirring stick (if not disposable then it has to be cleaned/washed immediately after mixing).
6. Pour the mixture directly into the molding cup. Make sure not to move the sample away from the center of the mold. Make sure to fill the mold completely with the liquid polymer.
7. Be sure to physically check the hardness of the epoxy base before you remove the sample from the mold.
8. Once the epoxy base has hardened, the sample can then be removed from the mold. The sample is now ready for grinding and polishing.

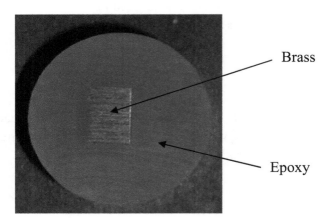

Figure 2. Mounted brass.

Stage 3: Grinding

Grinding of sample is necessary to produce a flat surface and remove any surface damage. Generally, several stages of grinding are required. Four grinding wheels are used; each wheel has a specific grit size. The first wheel is the roughest surface (240 grit) with surface roughness decreasing in succession. The last wheel has the least surface roughness (1000 grit).

Our sample will be ground using 120, 240, 400, 600, and 1000 grit abrasive disks one at a time.

Note: As the grit size increases, the surface of the grinding paper becomes smoother.

A. Rough Grinding

This is a corrective technique needed to remove gross surface irregularities. Additionally, protrusions or unwanted/excess polymeric material produced by the molding process are removed by grinding. Course grinding employs grinding using a 120 or 180 grit abrasive disk on the grinder wheel.

1. Turn on grinding wheel at low speed.
2. Turn on tap water onto grinding wheel. Don't apply too much water.
3. Grind the sample for about 2 minutes.
4. Turn the sample 90 degrees; grind the sample about 2 more minutes.
5. Make sure that the surface you are grinding remains flat.

Figure 3. Brass sample after the first stage of grinding. The image on the right shows the surface at 50X. Note it is difficult to have a focused view at this magnification due to deep scratches.

B. Fine Grinding

1. This step progressively removes coarse surface scratches utilizing a series of increasing grit size (240, 400, 600 and 1000) abrasive papers.
2. Between steps in the grinding process, the specimen should be thoroughly washed to prevent abrasive particles from being carried to finer grit papers. (Tap water is OK).
3. To perform fine grinding, manually hold the specimen steady against the rotating speed of the abrasive surface.
4. Each step should take approximately 2 minutes. After proper washing and drying, move to the next grit paper.
5. Rotate the specimen 90 degrees for this next step.
6. When the visible scratches from the previous step have been removed, continue grinding additional 6–8 strokes. See Figure 4 for the manual fine grinding of a brass sample.

Important tips about grinding:

1. Do not apply high pressure. Apply low and even pressure.
2. The surface must remain flat. Check for surface flatness regularly.

If this stage is not done properly, the remaining stages would be useless!

The comparison of the first and last grinding images give us a good idea of the improvement to the surface quality achieved thus far. This is shown in Figure 5.

Figure 4. The image above displays the brass sample after the final stage of grinding. The finish of the ground surface continues to become more precise. The surface becomes shinier too.

Figure 5. Comparison between the first stage of grinding and the final stage.

Stage 4: Polishing

Polishing of the sample produces **mirror-like** surface finish. The polishing process will remove the scratches produced by the grinding process. Generally multiple stages of polishing are required. Similar to grinding, 2–4 polishing wheels are required. Each wheel is covered with a cloth surface. Each surface is soaked with a mixture of abrasive particles (e.g., Alumina or diamond particles) of specific size and extender oil. The first wheel is soaked with 6-micron or 1-micron diameter particles. The last wheel is soaked with 0.1-micron diameter particles.

1. Deposit 6-micron or 1-micron diamond paste on the first polisher. Only a **very** small amount is needed.
2. Pour some extender onto the paste. A liquid extender is applied to aid in the distribution of diamond particles and provide lubrication.
3. Turn on the polishing wheel at low speed.
4. Polish the sample for about 2 minutes.
5. Turn the sample 90 degrees and polish for another 2 minutes.
6. Clean the sample **using distilled water.**
7. Deposit 0.1-micron diamond paste on the cloth located on the second polishing wheel.
8. Pour some diamond extender onto the paste.
9. Start the polishing process again at low speed till it looks shiny; see Figures 6, 7, and 8.

Polishing tips:

1. As with grinding, do not apply heavy pressure.
2. After each polishing stage rinse the sample with distilled water. This is mainly to avoid carrying abrasive particles from one wheel to another.
3. Always use the same abrasive mixture on one wheel to avoid abrasive cross-contamination.
4. The finished polished surface should be like a mirror, i.e., you should be able to see yourself clearly in it.

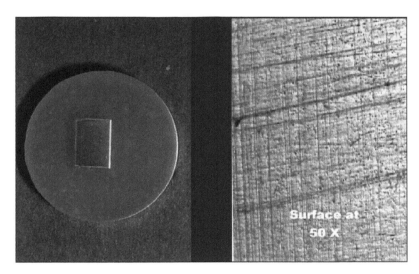

Figure 6. The image above displays the brass sample after the first stage of polishing. The image on the right shows the improvement that the polish process made.

Figure 7. The image above displays the brass sample after the final stage of polishing. The finish of the polished surface is mirror-like; you should avoid touching it to avoid scratching/contamination.

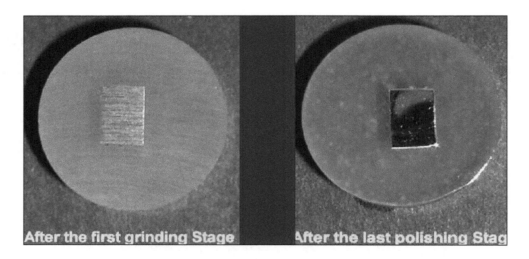

Figure 8. Side-by-side comparison between the first grinding stage and the last polishing stage provides a good idea of the improvement in surface quality.

Stage 5: Cleaning and Etching

Etching of the sample is performed to reveal the grain structure and other features of the sample. The surface of the sample is exposed to a corrosive chemical. The chemical attacks the surface and removes the weakly bonded atoms. The atoms at the grain boundaries are removed at a higher rate because the boundary atoms have a higher state of energy. As a result, the grain boundaries engrave deeply and look darker than the grains using a light microscope. Also, impurities react differently to the chemical and distinguish themselves in the image similarly.

An etchant is a mixture of chemicals such as acids, bases, alcohols, and water. Each metal is best etched by its own specific etchant. Materials handbooks (ASM Handbook of Crystallography) can be consulted for the type and make up of the etchants for different metals. The most suitable etchant is a diluted solution of several acids; see Table 1.

1. Wash your sample using distilled water.
2. Place your sample in the sonicator after filling one third of it with distilled water.
3. Leave the material sample in the sonicator for 45 sec.
4. Take the sample out from the sonicator and immediately wash with acetone.
5. Dry the sample with compressed air.
6. **Make sure your gloves are not torn. If they are, use a new pair.**
7. Pick out or prepare the appropriate etchant to use.
8. Pour droplets of the etchant over your sample—just enough to cover the material.
9. Start a stopwatch, or time using your watch, wait about 9 sec, then clean it using acetone.
10. The sample should be immediately dried well with air.
11. Do not touch the polished/etched specimen. This will alter the metal's surface condition.
12. The next step would be checking the material's grain size and boundaries through the use of a light microscope.

Table 1: Etchants for different metals.

1. Copper and Brass 2–3 sec	5 parts HNO_3	5 parts Acetic Acid	1 part H_3PO_4
2. Aluminum 60 sec	15 parts HF	10 parts H_3PO_4	60 parts H_2O
3. Steel (NITAL) 30–60 sec	5 parts HNO_3	95 parts Methanol	95 parts Ethanol

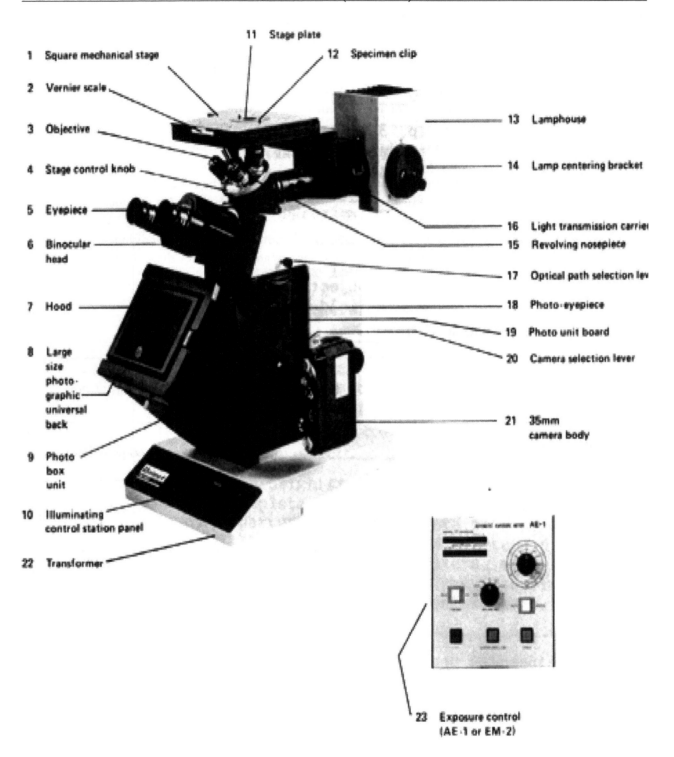

1 Square mechanical stage

2 Vernier scale

3 Objective

4 Stage control knob

5 Eyepiece

6 Binocular head

7 Hood

8 Large size photographic universal back

9 Photo box unit

10 Illuminating control station panel

22 Transformer

11 Stage plate

12 Specimen clip

13 Lamphouse

14 Lamp centering bracket

16 Light transmission carrier

15 Revolving nosepiece

17 Optical path selection lever

18 Photo-eyepiece

19 Photo unit board

20 Camera selection lever

21 35mm camera body

23 Exposure control (AE-1 or EM-2)

Settings on Automatic Exposure Meter

- 35mm (not 4x5)
- Shutter open (down)
- Auto (not manual)

Light Microscopy Procedure

1. Place your polished sample on the metallurgical microscope stage plate with the sample facing the beam of light directed from the apparatus.
2. Increase the light intensity by regulating it through using the illuminating control station panel.
3. Adjust the picture using the rough/fine knob regulator connected to the microscope to your right.
4. You must observe equiaxed grains for the sample used in this experiment (i.e., isotropic polycrystalline material).
5. A camera is connected to the microscope, which in turn is connected to a computer.
6. Pull out the optical path selection lever to switch the image to the camera.
7. Digital video pictures of the sample grains are captured and saved using Omnimet software.
8. Using the edit tool bar, 20 measurements of 20 different grains should be recorded using the circle option.
9. The average size of the aluminum grain is measured through calculating the mean diameter of the 20 diameters, and then the standard deviation is computed.
10. Each group must save the grain picture in their group folder on the desktop.
11. Students must save the image to a memory stick or a thumb drive so that it can be presented in their lab report.

Grain Size Determination

Estimate the average grain size using one of the following methods:

Method 1

By drawing circles (at least 20 circles per image) around different grains and finding the average diameter.

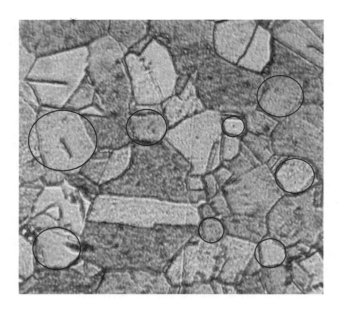

Figure 9. Image of a brass sample revealing the microstructure. Note that the grains are not true circles and hence the estimating circle can overestimate or underestimate the grain boundaries. As long as the overestimation and underestimation balance out, you then get a good estimate of the grain size.

Method 2

Using the line intercept method:
- Draw 10 arbitrary lines of equal length (say 100 mm) within the micrograph.
- Count the number of grains each line intercepts (N_1, N_2, N_3...N_{10}).
- Get average:

$$N_{avg} = \frac{N_1 + N_2 + ... + N_{10}}{10} \qquad (1)$$

Then find the grain size:

$$grain\ size = \frac{100\ mm}{N_{avg}} \qquad (2)$$

One useful objective of obtaining a micrograph is to determine the ASTM grain size number "n" (ASTM: American Society for Testing and Materials):

$$N = 2^{n-1} \qquad (3)$$

where N is the average number of grains per square inch (in²) at 100X magnification, and n is termed the ASTM grain size number.

If the magnification is other than 100X then use this alternative formula:

$$N_M \left(\frac{M}{100} \right)^2 = 2^{n-1} \qquad (4)$$

where N_M is the number of grains per in^2 at a magnification of MX.

The grain size number gives you information about the relative strength of the material. The larger the value of n, the more grains exist per square inch. The more grains exist, the stronger the metal (for a given material with a fixed thermal and deformation history).

The first step in finding the ASTM grain size number is to count the number of grains in the photograph. The grains cut by the edge of the picture count as half grains.

For 100X, we have from equation (3):

$$n = \frac{\log N}{\log 2} + 1 \qquad (5)$$

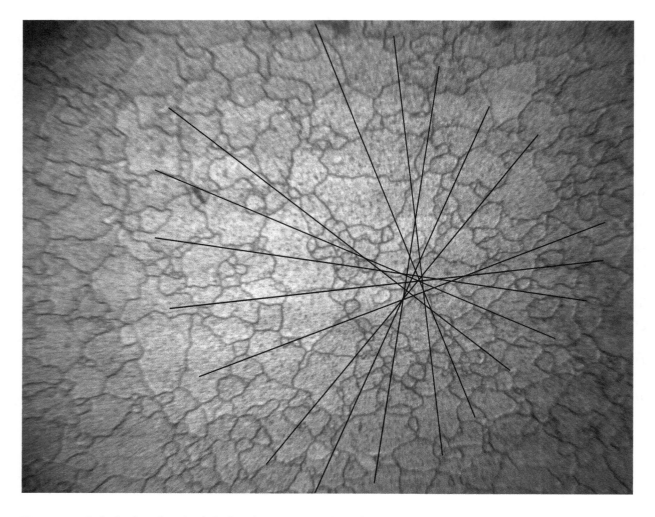

Figure 10. Polished and etched 6061 Aluminum at 60X showing multiple intercept lines.

TO DO LIST

1. Discuss possible reasons if you don't get good results (images).
2. If you skip 600 and 1000 grit grinding, and jump directly from 400 grit size to polishing, it is still likely you will get mirror-like smooth surface eventually. What will go wrong in this case regarding the microstructure you observe?
3. Choose one of your "good" images for aluminum/brass **at an arbitrary magnification** and find the ASTM grain size number. Compare your results to listed ASTM grain size numbers for the sample.
4. Does the grain size number increase or decrease with decreasing grain size? Why?

Note to Instructors: You can augment or edit the above procedure or To-Do List in any way that is most suitable for your lab/class situation. If you look into the literature, or even do an online search, there are multiple chemical etchants that can be used for the same material (e.g., one can use 5g of ferric chloride, 96ml of ethyl alcohol, and 2ml hydrochloric acid (for 1 second to several minutes) to etch copper, aluminum, magnesium, nickel, and zinc alloys). This is somewhat of an art rather than a science; trial and error is sometimes necessary to achieve best results. One of the most important things about microphotography using a microscope, besides having the actual equipment, is the obtained images need to be scaled. The way this is done is to calibrate the objectives and images taken by them using a linear scale of some sort. Typically this scale is a transparent glass ruler with millimeter (sub-millimeter) markers inscribed on it.

CHAPTER 6
HARDNESS MEASUREMENTS (EXPERIMENT/LAB 4)

T he purpose of this lab is to learn about some different hardness measurement techniques that commonly exist in practice.

INTRODUCTION

Hardness is the ability of a material to resist localized plastic deformation, i.e., indentation. Hardness provides a measure of a material's strength (in many cases hardness values are linearly proportional to material strength values). It can easily be measured in a nondestructive manner (without actually breaking the material tested). Hard materials are strong and offer more resistance to plastic deformation. They also resist wear during contact. When manufacturing gears, it is critical that the surface hardness (of contacting gear teeth) be very high in order to reduce wear during operation. Excessive wear could cause increased noise and chatter, leading to eventual failure of the gear mechanism.

Many measurement scales are based upon some type of standard. The freezing and boiling points of water determine temperature scales. Linear measurement is based upon the standard of inches and meters. Weight is based on a given volume of water. Unlike these previous examples, hardness has no absolute scale.

The hardness of the material is determined by indenting its surface using an indenter that has the shape of a ball, a pyramid, or a cone. Hardness is then calculated based on the value of the applied load, cross-sectional area of the indentation spot, or the depth of indentation. The indenter is generally made from hardened steel, tungsten carbide, or diamond.

High hardness means:
- Resistance to plastic deformation or cracking
- Better wear properties
- High material strength

Several hardness-testing techniques have been developed such as Rockwell, Brinell, Vickers, and Knoop (see the sections below for details on each method). In all these techniques, a small indenter is forced into the surface of a material to be tested, under controlled conditions of load, time of application, and rate of application. The depth or size of the resulting indentation is measured, which in turn is related to a hardness number. The softer the material, the larger the indentation size, the bigger the indentation depth, and the lower the hardness index number. Measured hardnesses are only relative (rather than absolute), and care should be exercised when comparing values determined by different techniques.

Test	Indenter	Side View of indentation	Top View of indentation	Load	Hardness Number Formula
Brinell	10 mm (or 5 mm) sphere of steel or tungsten carbide	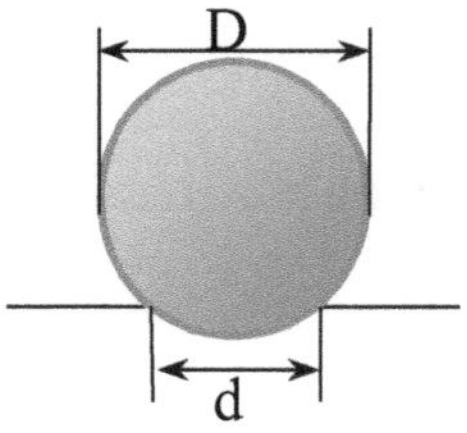	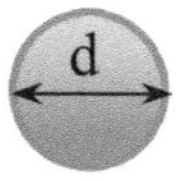	P	$HB = \dfrac{2P}{\pi D[D - \sqrt{D^2 - d^2}]}$
Knoop micro-hardness	Diamond pyramid	 $l/b = 7.114$	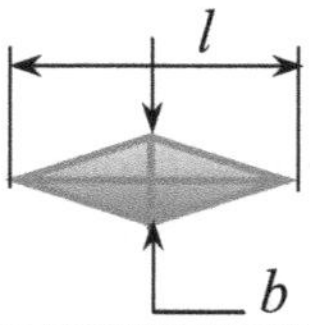	P	$HK = 14.2P / l^2$
Vickers micro-hardness	Diamond pyramid	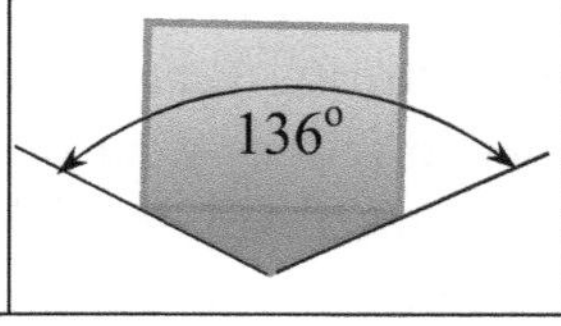 $136°$	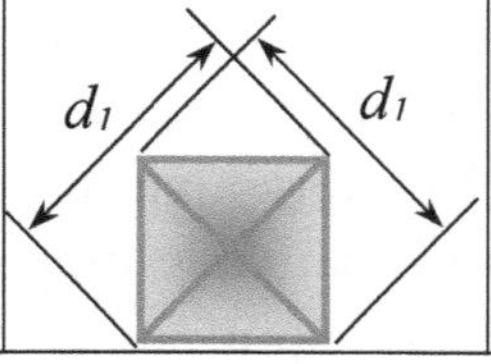	P	$HV = 1.854P / d_1^2$

Figure 1. Some common hardness measurement techniques. P **(the applied load) here is in kg. Also,** D, d, d_1 **and** l **are all in mm.**

A. ROCKWELL HARDNESS TEST

The Rockwell hardness test is an empirical indentation hardness test. It possesses many advantages over the other hardness tests. The test is fast, inexpensive, and relatively non-destructive (in the artificial Rockwell range mostly), leaving only a small indentation in the material.

Besides the hardness number, the Rockwell hardness test can produce important information about metallic materials, such as the ultimate strength, wear resistance, and ductility. The test is very practical for material selection in process and quality control of commercial products. Therefore, it is not unusual to find a Rockwell hardness machine in almost every testing environment, from industrial facilities to environmentally controlled academic laboratories.

Rockwell Indentation Test Principle

Most indentation hardness tests are a measure of the indentation residue upon the removal of the indenter tip away from the material undergoing hardness testing. As there are limitations for materials in terms of size and mechanical properties, there is a wide range of different indenter tips that vary in size, shape, and sometimes tip material. In the case of the Rockwell hardness test, two levels of force are applied to the indenter at specified rates and with specified dwell times, as illustrated for the Rockwell C scale (HRC) test in Figure 2.

Unlike the Brinell and Vickers/Knoop hardness tests, where the size of the indentation area is measured following the indentation process, the Rockwell hardness of the material is based on the difference in the indentation depth at two specific times during the testing cycle, indicated by the X marks in Figure 2. The value of hardness is calculated using a formula (see below) that yields a dimensionless number falling within an arbitrarily defined range of numbers known as a Rockwell hardness scale.

The general Rockwell test procedure is the same, regardless of the Rockwell scale or indenter being used. The indenter is brought into contact with the surface of the material to be tested, and a preliminary force (referred to as the minor load) is applied to the indenter. The preliminary force is usually held constant for a set period of time (dwell time), after which the depth of indentation is measured. After the measurement is made, an additional amount of force is applied at a set rate to increase the applied force to the total force level (referred to as the major load). The total force is held constant for a set time period, after which the additional force is removed, returning to the preliminary force level. After holding the preliminary force constant for a set time period, the depth of indentation is measured for a second time, followed by removal of the indenter from the test material. The measured difference between the first and second indentation depth measurements, h, (see Figure 2) is then used to calculate the Rockwell hardness number. In summary, the Rockwell Hardness number is based upon the difference in the depths to which an indenter is driven by a definite light or "minor" load and a definite heavy or "major" load.

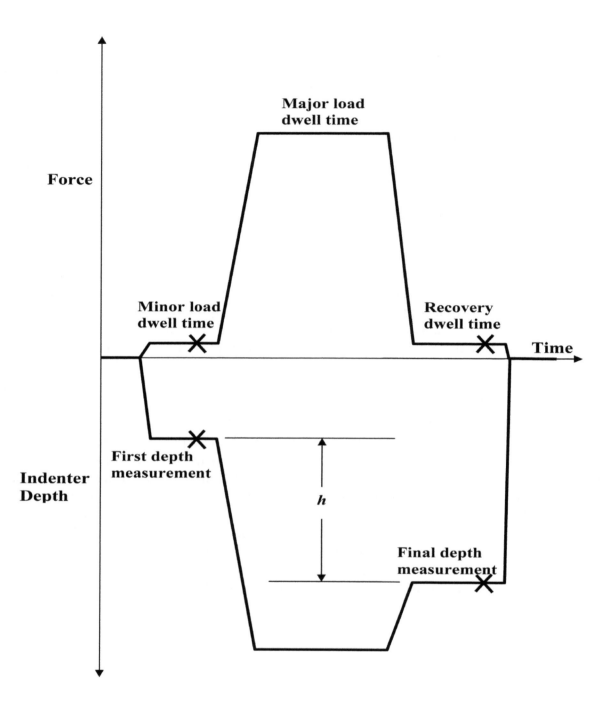

Figure 2. Plots of force vs. time (top) and indenter-depth vs. time (bottom) for an HRC test illustrating the testing cycle parts and the difference in indenter depth measurements _h_.

ROCKWELL HARDNESS SCALES

Many manufactured products are made of different types of metals and alloys varying in hardness, size, and thickness. To accommodate the testing of these diverse products, several different indenter types were developed for the Rockwell test to be used in conjunction with a range of standard force levels. Each combination of indenter type and applied force levels has been designated as a distinct Rockwell hardness scale. ASTM (American Society for Testing and Materials) defines thirty different Rockwell scales. Table 1 shows some of them.

Rockwell hardness scales are divided into two categories: Regular Rockwell scales (R) and Superficial Rockwell scales (S). Both categories use the same types of indenters. The regular Rockwell scales employ higher force levels. For these scales, the preliminary force level is 98.07 N (10 kgf), and the standard total force levels ranges from 588.4 N (60 kgf) to 1471 N (150 kgf). The superficial Rockwell scales employ lighter force levels, and are typically for use on thinner materials. For the superficial Rockwell scales, the preliminary force level is 29.42 N (3 kgf), and the standard total force levels range from 147.1 N (15 kgf) to 441.3 N (45 kgf). Table 2 provides typical applications for the different Rockwell scales as recommended by ASTM, and it lists the appropriate type of indenter and force levels to be used with the particular scale.

Rockwell Hardness Number

A Rockwell hardness measurement is reported as a Rockwell hardness number, <u>without units</u>. The Rockwell hardness number is calculated from the difference in the indentation depths before and after application of the total force, while maintaining the preliminary test force. The difference in indentation depths is measured as h (see before). The calculation of the Rockwell hardness number is dependent on the specific combination of indenter type and the forces that are used.

— For scales that use a spheroconical diamond indenter, the Rockwell hardness number is calculated from h (in mm) as follows:

$$Regular\ Rockwell\ Hardness = 100 - \frac{h}{0.002\ mm} \tag{1}$$

$$Rockwell\ Superficial\ Hardness = 100 - \frac{h}{0.001\ mm} \tag{2}$$

— For scales that use a ball indenter, the Rockwell hardness number is calculated from h (in mm) as:

$$Regular\ Rockwell\ Hardness = 130 - \frac{h}{0.002\ mm} \tag{3}$$

$$Rockwell\ Superficial\ Hardness = 100 - \frac{h}{0.001\ mm} \tag{4}$$

	Scale Symbol	Indenter Type	Minor Load N (kgf)	Major Load N (kgf)	Typical Applications
Regular Rockwell Scales	A	Spheroconical Diamond	98.07 (10)	588.4 (60)	Cemented carbides, thin steel, and shallow case hardened steel
	B	Ball, 1/16 in	98.07 (10)	980.7 (100)	Copper alloys, soft steels, aluminum alloys, malleable iron, etc.
	C	Spheroconical Diamond	98.07 (10)	1471 (150)	Steel, hard cast irons, pearlitic malleable iron, titanium, deep case hardened steel
	D	Spheroconical Diamond	98.07 (10)	980.7 (100)	Thin steel and medium case hardened steel, pearlitic malleable iron
	E	Ball, 1/8 in	98.07 (10)	980.7 (100)	Cast iron, aluminum and magnesium alloys, and bearing metals
	F	Ball, 1/16 in	98.07 (10)	588.4 (60)	Annealed copper alloys, and thin soft sheet metals
	G	Ball, 1/16 in	98.07 (10)	1471 (150)	Malleable irons, copper-nickel-zinc alloys
	H	Ball, 1/8 in	98.07 (10)	588.4 (60)	Aluminum, zinc, and lead
	K	Ball, 1/8 in	98.07 (10)	1471 (150)	Bearing metals and other very thin or soft materials
Superficial Rockwell Scales	15N	Spheroconical Diamond	29.42 (3)	147.1 (15)	Similar to A, C, and D scales, but for thinner gauge material or case depth
	30N	Spheroconical Diamond	29.42 (3)	294.2 (30)	
	45N	Spheroconical Diamond	29.42 (3)	441.3 (45)	
	15T	Ball, 1/16 in	29.42 (3)	147.1 (15)	Similar to B, F, and G scales, but for thinner gauge material
	30T	Ball, 1/16 in	29.42 (3)	294.2 (30)	
	45T	Ball, 1/16 in	29.42 (3)	441.3 (45)	

Table 1. Rockwell hardness scales with the corresponding indenter type, applied forces, and typical applications.

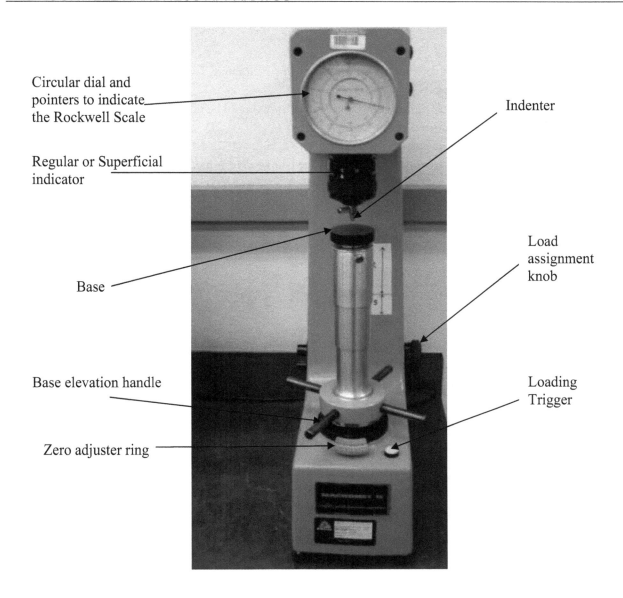

Circular dial and pointers to indicate the Rockwell Scale

Indenter

Regular or Superficial indicator

Load assignment knob

Base

Base elevation handle

Loading Trigger

Zero adjuster ring

Figure 3. Rockwell Hardness Tester.

Rockwell Hardness system uses a direct readout machine determining the hardness number based upon the depth of penetration of either a diamond cone or a steel ball. The ball indenters are chucks that are made to hold 1/16 in or 1/8 in diameter hardened steel balls. Also available are ¼ in and ½ in ball indenters for the testing of softer materials. There are two types of anvils that are used on the Rockwell hardness testers. The flat faceplate models are used for flat specimens. The "V" type anvils hold round specimens firmly.

Test blocks or calibration blocks are flat steel or brass blocks, which have been tested and marked with the scale and Rockwell number. They should be used to check the accuracy and calibration of the tester frequently.

Test Setup

1. Select the test block, which corresponds to the scale being calibrated. Two reference samples will be tested first using 30T and B scales.
2. Place the test block on the anvil beneath the indenter. Care should be taken so that the test block is clean, smooth, and properly supported by the anvil.
3. Determine the material scale written on the sample.
4. Using the table on the machine (shown in the image below) determine:

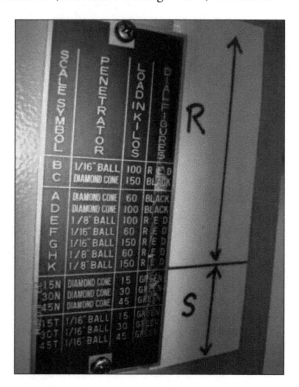

a. Appropriate probe used. If you need to change the indenter, unscrew the pin attached to it, change the indenter, and screw the pin back.

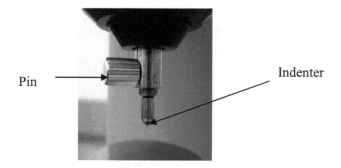

Pin   Indenter

b. Applied load on the specified material scale; and set the load through the knob attached to the tester to your right.

c. Type of loads applied—Superficial Rockwell (S), or Regular Rockwell (R) loads; and set it by rotating the switch that is located above the indenter to S or R.

d. The color of the scale used to measure the hardness using the hardness meter.
5. Now the reference sample is ready to be tested.

6. Turn the hand wheel clockwise to raise the anvil assembly. When contact is made between the indenter and the test block, the dial pointer will move in a clockwise direction.
7. Continue turning the hand wheel until the probe is in contact with the sample; until the small indicator on the hardness meter is zeroed (indicator pointing to the red dot). The minor load has now been applied.
8. Turn the AC switch on.
9. Press the white button (Loading Trigger) located under the wheeler to start running the hardness test.
10. Note the movement of the large pointer after releasing; allow the pointer to stabilize. This may take a few seconds.

11. Read the hardness meter to determine the material's hardness and record it. Compare the recorded hardness number to the number on the reference sample; they should be close. If readings are consistently

Loading Trigger

higher than what is marked on the test block, minor adjustment is needed. Continue the calibration procedure until the tester is reading within the limits of the test block.

12. Always take two (2) or more readings with a test block before using the tester. **<u>DON'T indent close to an existing indentation.</u>**

13. **<u>Do not use both sides of the test block.</u>**

14. Remove the reference sample after you are done with this setup.

Making a Test

15. The same procedure is used to test Al 2024-T351 sample, and the specimen is indented 20 times.

16. Another sample 4340 steel is tested by performing 20 indentations as well.

17. Perform these two tests using the A scale (each student performing two tests for a total of two hardness measurements).

18. We end the experiment by comparing aluminum to steel hardness.

Trial	Al (2024-T351)	Steel (4340)
1		
2		
3		
4		
5		
6		
7		
8		
9		
10		
11		
12		
13		
14		
15		
16		
17		
18		
19		
20		
Average Value		

B. BRINELL HARDNESS TEST

We have introduced the Rockwell hardness test as a common technique to determine the hardness of a material. The Brinell hardness test is another common technique for hardness measurement. Rockwell hardness testing is generally used on harder steels and samples where the Brinell hardness test leaves too large an impression on the specimen, hence rendering it many times undesirable.

The Brinell testing machine is relatively simple. The large indenter averages out load variations in the specimen, and it can be used on relatively rough surfaces. However, it does leave rather large indentations in the test specimens. The large Brinell indenter will not make a significant impression in hard materials, so the test is not useful beyond the Rockwell C-60 range. The thickness of the specimen being tested should be about 10 times the depth of the indentation for best accuracy. This prevents the test from being conducted on thin materials. In practice, good values may be obtained if there is no visible effect on the back of the specimen. A similar restriction applies to indenting close to edges. Ideally, an indent should not be made within 4–5 indentation diameters from the edge of a specimen.

A close correlation between the Rockwell and Brinell hardness numbers has been developed. Conversion charts/tables/online calculators are available for changing from one to the other. The Brinell hardness number is also directly related to the strength of a material. Hence, hardness testing can replace slower tensile testing to determine, for example, the strength of steel.

In Brinell tests, a hard (hardened steel or tungsten carbide), spherical indenter is forced into the surface of the metal to be tested; see Figure 4. The diameter of indenter (D) is 10.00 mm (or 5.00 mm). Standard loads (P) range between 500 and 3000 kg (in 250-kg or 500-kg increments); during a test, the load is maintained constant for a specified time (between 10 and 30 sec). Harder materials require greater applied loads. The Brinell hardness number, HB, is a function of both the magnitude of the load and the diameter of the resulting indentation:

$$HB = \frac{2P}{\pi D[D - \sqrt{D^2 - d^2}]} \tag{5}$$

This diameter is measured with a special low-power microscope, utilizing a scale that is etched on the eyepiece. The measured diameter is then converted to the appropriate HB number using a chart or the formula above. Only one scale is employed with this technique.

Maximum specimen thickness as well as indentation position (relative to specimen edges) and minimum indentation spacing requirements are the same as for Rockwell tests. In addition, a well-defined indentation is required; this necessitates a smooth flat surface in which the indentation is to be made.

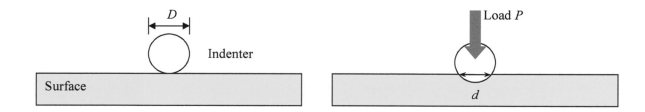

Figure 4. Brinell Testing Technique

Brinell Hardness Apparatus

The Dyna Brinell apparatus is shown in Figure 5.

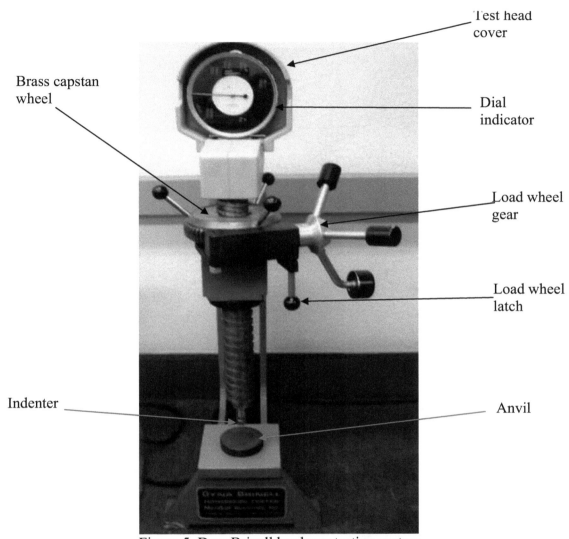

Figure 5. DynaBrinell hardness testing system.

Figure 5. DynaBrinell hardness testing system.

Brinell Test Procedure:

1. The needle on the dial indicator should rest on zero. If adjustment is needed turn the faceplate of the dial indicator until the needle is zeroed.

2. Place the specimen to be tested onto the anvil.

3. To bring the indenter into contact with the workpiece, release the load wheel latch and disengage the load wheel gear by swinging it away from the brass capstan wheel.
4. Turn the brass capstan wheel to set the indenter ball lightly into the specimen.

5. Re-engage the load wheel and close the latch. If the gears do not mesh, simply turn the load wheel a bit until they do fit.
6. Turn the load wheel **clockwise** until the needle on the dial indicator registers that the desired load has been applied. This position should be maintained for 10–30 seconds.
7. Remove the load and back the indenter away from the work piece.
8. Release the load wheel latch and turn the capstan wheel counterclockwise until the specimen is free.
9. After the indentation is made, it is read (in the normal Brinell method) using a Brinell scope. A portable 7X scope is provided; Figure 6.

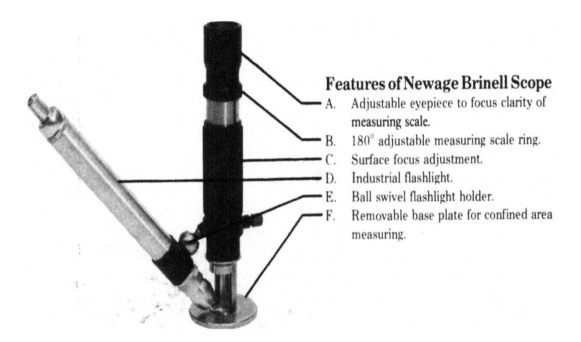

Features of Newage Brinell Scope

A. Adjustable eyepiece to focus clarity of measuring scale.
B. 180° adjustable measuring scale ring.
C. Surface focus adjustment.
D. Industrial flashlight.
E. Ball swivel flashlight holder.
F. Removable base plate for confined area measuring.

Figure 6. Brinell scope.

10. When measuring an impression, determine if the removable base (F) may hinder the reading. If so, unscrew plate. Place the scope over the impression and illuminate the impression using the flashlight (D). Adjust the ball swivel holder (E) so the light beam falls on the impression you are measuring.
11. Bring the impression in focus by adjusting the widest ring (C). Then bring the measuring scale into focus by adjusting eyepiece (A). Now adjust the measuring scale ring (B) for alignment of scale across the impression. Adjust illumination for proper lighting with ball swivel holder.
12. Place zero measuring scale on the left edge of the impression at its widest point. Now measure across the impression to right edge to determine the diameter in mm.
13. Read the distance across the impression by reading in millimeters and tenth of millimeters. If the right side of the impression falls halfway between two lines on the scale, it is considered a half of a tenth of a millimeter or 0.05 mm. It is important to take the most accurate measurement possible. Ask each member of your team to read the impression diameter separately.
14. Once the measurement is made, convert to a Brinell hardness number using (a) the equation provided in the lab manual, or (b) Brinell chart (be sure to read from the columns corresponding to the load and ball size actually used).

15. When making more than one indentation using the same sample, the distance between the center of an indentation from edge of the sample or edge of another indentation should be at least 2.5 times the diameter of the indentation.
16. The thickness of the piece should be at least 10 times the depth of the indentation.

Now determine the Brinell hardness number (HB) for a reference metal sample.

Trial	D (mm)	Readings of Depression Diameter (d), mm	Avg. Dia. (d), mm	Brinell Number
1				
2				

C: MICRO-INDENTATION: VICKERS HARDNESS TEST

Introduction:

Micro-indentation hardness testing (MHT) that includes Knoop, Vickers, and Berkovich tests provide the same valuable information obtained via bulk hardness testing (Rockwell and Brinell) but with a much smaller indent and thus they are sometimes referred to as non-destructive mechanical tests. Because the indents are small, MHT can be used for many parts or material forms that are too small or too thin to test with bulk test procedures. This includes very thin films used in MicroElectroMechanical (MEMS) applications or for very soft materials (polymers). Furthermore, due to the small size of the indent, it is possible to use MHT to measure the hardness of different constituents composing a microstructure (different phases or different inter-phases such as fibers and matrix). For example, the determination of hardness of specific types of carbides, nitrides, borides, sulfides, or oxides in metals has been widely performed, particularly in wear and in machinability research. There is a long list of applications where MHT is essential.

A few examples are hardness testing of thin materials, measurements of case hardening, and measuring the hardness of different phases in a microstructure. For this last application, very small probing tools of the different phases are needed since the phases would typically range in the micron range. It is because of this that MHT is a valuable tool to characterize the mechanical properties since the area to be probed is so small. More recently, nano-indentation testing has been replacing micro-indentation testing for the mechanical characterization, through hardness readings, of small microstructural features within a material.

In MHT, a diamond indenter of specific geometry is pressed into the surface of the test specimen using a known applied load of 1 to 1000 gf. There are different protocols with different ranges of the applied force for Vickers hardness test. For example, ASTM E 384 identifies the load range from 1 to 1000 gf, and this is the commonly accepted range in the United States.

At any case, the Vickers hardness number relies on measurements made of the indent formed in the surface of the test specimen. It is important to point out that the elastic recovery is assumed to not occur upon breaking the indenter/sample contact, but this is rarely the case. The Knoop test is claimed to eliminate recovery, but again, this is not true for tests of metallic materials. The Berkovich indentation relies heavily on the elastic recovery to estimate the hardness.

The imprint left on the sample by a Vickers indenter is a diamond shape with two diagonals. Both diagonals are measured and the average value is used to compute the Vickers hardness (HV). The hardness number is actually based on the surface area of the indent itself divided by the applied force, giving hardness units of kgf/mm² or gf/mm². In the Knoop test, only the long diagonal is measured, and the Knoop hardness (HK) is calculated based on the projected area of the indent divided by the applied force, also giving test units of kgf/mm² or gf/mm². In practice, the test units kgf/mm² (or gf/mm²) are not reported with the hardness value.

Vickers Hardness Test

The Vickers hardness test operates on similar principles to the Brinell test, the major difference being the use of a square-based pyramidal diamond indenter rather than a hardened steel ball. The test was developed because the Brinell test, using a spherical hardened steel indenter, could not test hard steels, as spherical indentation generally produce mostly elastic strains. Furthermore, the diamond does not deform at high loads (very hard and very stiff) so the results on very hard materials are more reliable. Another major reason for using Vickers/Knoop over Brinell is the much less intrusive nature of this microhardness test. Figure 7 shows examples of Vickers indents to illustrate the influence of the magnitude of test force on indent size.

Figure 7. Vickers indents made with different applied loads.

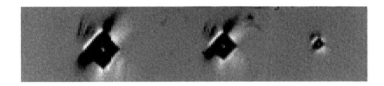

The pyramidal shape with an angle of 136° between opposite faces was chosen in order to obtain hardness numbers that would be as close as possible to Brinell hardness numbers for the same specimens. Unlike Rockwell tests, the Vickers and Brinell tests have the great advantage of using one hardness scale each to test all materials.

In the Vickers test, the force is applied smoothly, without impact, and held in contact for 10 to 15 sec (but can be 20 sec). The force must be known precisely. After the force is removed, diagonals, d_1 and d_2 (illustrated in Figure 8) are measured and their average (d) is used to calculate the surface area. Dividing the applied load by the calculated area will give the Vickers hardness number (HV) according to:

$$HV = \frac{2000P\sin\left(\alpha/2\right)}{d^2} = \frac{1854.4P}{d^2} \tag{6}$$

where d is the mean diagonal in mm, P is the applied load in gf, and α is the face angle (136°).

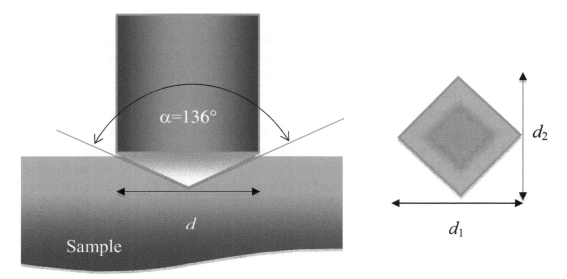

Figure 8. Vickers hardness test.

Most modern MHT units have the calculation capability built in and display the hardness value along with the measured diagonals. Alternatively, older testers come with a book of tables of HV as a function of P and d.

The hardness may also be reported as Vickers hardness number (VHN) or Diamond Pyramid Number (DPN). A Vickers indentation is smaller than a Brinell impression. A small impression demands the sample surface to be flat and perpendicular to the indenter and should have a better than 300 grit finish. Thus some grinding and polishing before a Vickers test are highly recommended.

Errors in Hardness Testing

There are many factors that can affect the accuracy of the hardness test. Some of these such as flatness and surface roughness. Ideally, the indented surface should be flat within ± 1°I [4]. This is why grinding and polishing are highly recommended.

The correct load must be applied and there must be no friction in the loading system, otherwise the impression will be smaller than expected, yielding a much higher VHV number than the sample's actual property. Thus, regular maintenance and calibration of the machine is essential.

The condition of the indenter tip is crucial—whilst the Vickers diamond is unlikely to deteriorate with use unless it is damaged or loosened in its mounting by clumsy handling, the Brinell metal ball will deform over a period of time and inaccurate readings will result. And for any hardness test, the duration of the application of load is important and must be controlled.

The specimen geometry is also important—if the specimen is too thin, the hardness of the specimen support underneath will affect the results. As a rule of thumb the specimen thickness should be ten times the depth of the impression for the Brinell test and twice that of the Vickers diagonal [4]. Similarly, if the impression is too close to the specimen edge then low hardness values will be recorded —again, as a general rule the impression should be 4 to 5 times the impression diameter from any free edge[4].

Vickers Hardness Tester

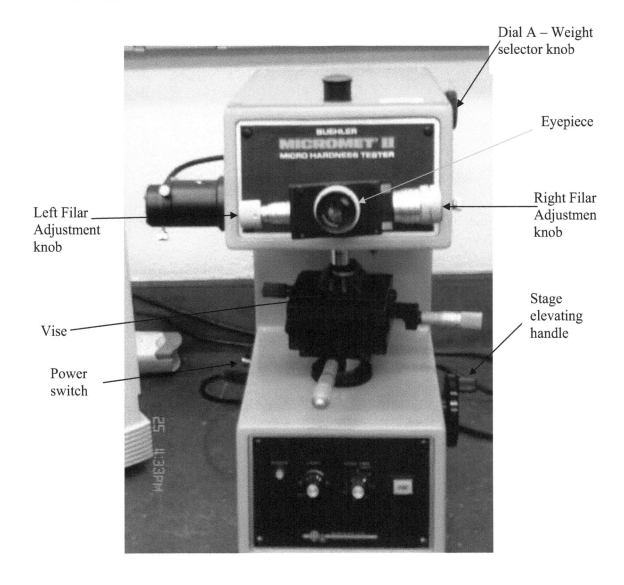

Dial A – Weight selector knob

Eyepiece

Left Filar Adjustment knob

Right Filar Adjustmen knob

Vise

Stage elevating handle

Power switch

The Vickers Hardness Microhardness Tester Procedure

A: Indentation

1. Place the specimen to be tested in the vise so that the surface to be tested is perpendicular to the diamond indenter.

2. Turn the power switch on (toggle switch on the left side of the instrument). The power indicator will illuminate on the front panel.

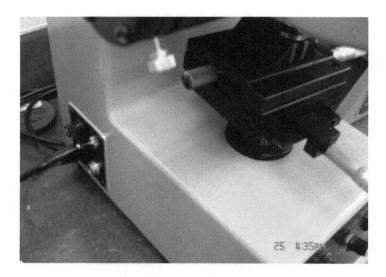

3. Rotate the Objective-Indenter turret to the 10X scanning objective. The Turret is equipped with click stops to assist with proper alignment.

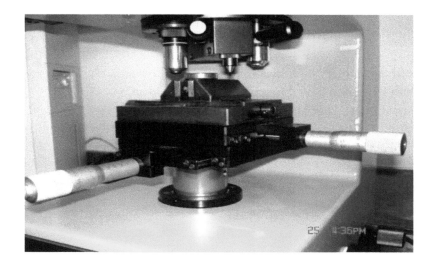

4. The specimen can be brought into focus by the Stage Elevating Handle located on the right side of the tester.
5. The illuminator intensity control knob, located on the front panel, can be used to vary the light brightness.
6. Select the area of the sample to be tested.
7. Rotate the Objective-Indenter Turret to the 40X scanning objective and re-focus if necessary.
8. Select the weight for the load application. The Dial A – Weight selector knob is located on the right side of the tester.
9. Set the load timer for the length of time that the load is to be applied (usually 10 sec).
10. Rotate the Objective-Indenter Turret to the indenter and press the START Button. The START button will remain illuminated for the duration of the testing cycle. When the start button light goes off, the test is done.

B: Measurement

1. Rotate the Objective-Indenter Turret back to the 40X objective.
2. If the stage has not been moved, the specimen should be in focus.
3. Set the micrometer (on your right-hand side) to zero.
4. Advance the Left Filar Line (controlled by the adjustment knob to the left of the Eyepiece) so that the inner edge of the line just touches the left-most point of the impression.

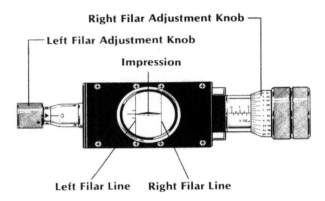

5. Advance the Right Filar Line (controlled by the adjustment knob to the right of the Eyepiece) so that the inner edge of the line just touches the right-most point of the impression.
6. Measurement is achieved by touching the inside of both Filar lines to the outermost points of the indentation.
7. Record the value of the Filar Micrometer inscribed in both the barrel and the Right Filar Adjustment Knob. The micrometer is calibrated in 0.5 micron increments (the smallest division on the drum) with one complete revolution of the drum being equal to 25 microns. The value of the micrometer reading is equal to the width of the impression made by the indenter.

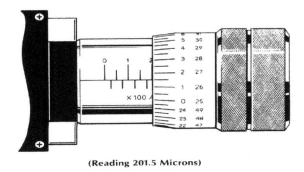

(Reading 201.5 Microns)

8. Perform the indentation procedure above twice—once for once for (500 g) load and again for (1000 g) load. Repeat.

Load	Reading of Depression dimension, mm		Avg. Diag.	Vickers number
	d_1	d_2		
500 g				
500 g				
1000 g				
1000 g				

TO-DO LIST

1. Find the average and standard deviation of the Rockwell hardness number of both materials tested. Comment on the results and give reasons for large deviations (if any).
2. Compare the results obtained for these two materials with those referenced in materials handbooks.
3. What are the advantages and disadvantages of Rockwell hardness testing?
4. What happens if the indentation is too close to previous indentation? (Hint: investigate how cold-working or strain-hardening affects hardness.)
5. Using the average of 3 to 4 diameter measurements, find the Brinell hardness number of the specimen at two different locations. Compare your results to the HB printed on the reference sample. Explain any discrepancy between your computed value and the reference value.
6. Find the Vickers hardness number for the reference sample using 500 g and 1000 g loads (twice each).
7. List some advantages of the Vickers hardness test over the Brinell test.
8. Discuss the sample thickness requirement for the three tests Rockwell, Brinell, and Vickers.

Note to Instructors: You can augment or edit the above procedure or To-Do List in any way that is most suitable for your lab/class situation. The specific indentation machine use will vary. Some come in digital form for example. The important thing for the Brinell and Vickers/Knoop hardness tests is to know the amount of load being applied to the surface. This can be done via a load cell of some sort. Another important thing is to be able to use the right indentation tip shape/size and be able to measure the indentation spot size with either magnifying glass or a microscope. In the case of Brinell hardness, the spot is normally big enough that it can be measured to a good accuracy with a magnifying glass coupled with a Vernier caliper. Another important thing in hardness testing is to use the same parameters (e.g., load-application time) between measurements that are to be compared later. It is also very important to stay away from sample edges or other indentation spots when doing indentation. The last important thing is to avoid indentation depths that are close to the bottom of the substrate that you are indenting. If that means to use a lower indentation force ,then so be it.

CHAPTER 7
POWDER METALLURGY (EXPERIMENT/LAB 5)

The students will become familiar with the process of powder metallurgy (PM) through compacting zinc or other metal powder into a solid tablet under varying pressures, followed by hardness testing of these tablets.

INTRODUCTION

Powder metallurgy is a forming and fabrication technique consisting of three major processing stages. First, the primary material is physically powdered, divided into many small individual particles (microns or nanometers in size). Next, the powder is injected into a mold or passed through a die to produce a weakly cohesive structure (via cold welding) very near the dimensions of the object ultimately to be manufactured. Pressures of 10–50 tons per square inch are commonly used. Also, to attain the same compression ratio across more complex pieces, it is often necessary to use lower punches as well as an upper punch to obtain uniform density. Finally, the end part is formed by applying pressure, high temperature, long setting times (during which self-welding occurs), or any combination thereof.

In powder metallurgy or ceramics, it is possible to fabricate components that otherwise would decompose or disintegrate. All considerations of solid–liquid phase changes can be ignored, so powder processes are more flexible than casting, extrusion, or forging techniques. Controllable characteristics of products prepared using various powder technologies include mechanical, magnetic, and other unconventional properties of such materials as porous solids, aggregates, and intermetallic compounds. Competitive characteristics of manufacturing processing (e.g.,, tool wear, complexity, or vendor options) also may be closely regulated.

Today, PM products are used in a wide range of industries, from automotive and aerospace applications to power tools and household appliances.

In our experiment, we will skip the third step of heat-treating the newly formed solid and go directly into hardness testing and microscopy.

Equipment/Hardware and Materials

- Particle masks
- Safety glasses
- Latex gloves
- Electronic 500g scale, or any other accurate balance (within 0.01g or better)
- Plastic tray for powder holding and mixing
- Powder scoop
- Wooden or metal dowel
- Metal file
- Instron Universal Tension/Compression Tester (or any other loading frame with a loading cell). See Figure 1.

Figure 1. A compression-capable machine to compact the powder.

- Instron grip for steel plunger/punch
- Steel pin for holding grip in place in the machine
- Cylindrical die
- 4340 steel plunger
- Steel discs (two count)

Figure 2. (From left to right) A cylindrical steel die, a steel plunger/punch, two steel discs for compacting powder in between, and a grip to hold the plunger and attach it to the machine for axial compression.

- Zinc or other metallic powder
- Vickers hardness tester
- Microscope
- Digital caliper

PROCEDURE

Powder metallurgy:

1. Prepare the electronic scale by powering it on a few minutes before use. The load cells need to be "warmed up" before giving accurate readings. Use this time to put on all safety gear including goggles, masks, and gloves.
2. Using the scale, powder scoop, and powder tray, measure 2 grams of zinc powder. Account for the weight of the tray by zeroing the scale with just the tray on it.
3. Use wooden dowel to flatten out any powder that may have clumped together into observable balls or chunks.
4. Insert one metal disc into the die's cylindrical cavity. Let the disc set flush with the bottom of the die, then pour powder into the cylindrical mold, on top of the first metal disc. Then put the second metal disc on top of the powder. (See diagram for clarification.)

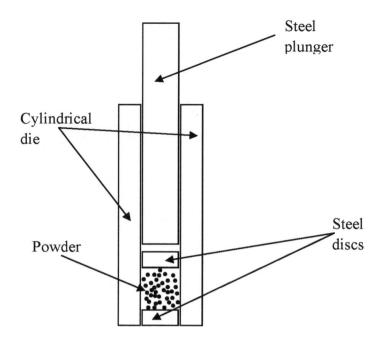

Figure 3. The different parts used in the powder compaction process (most of them shown in Figure 2).

5. Set up the Instron tension/compression tester for pressing the powder. This includes locking the plunger grip into place with the steel pin, screwing the steel plunger into the grip, and adjusting the stage to the proper height.

Figure 4. The assembly (different parts) used in the compaction process as it is attached to a grip in the compression machine. Note the steel pin holding the grip into place in the machine.

6. Place mold into position under the plunger. Make sure the plunger is aligned vertically into the mold to the best of your ability. This is a VERY IMPORTANT step as misalignment can cause damage to the parts used.

7. Start Instron software and configure settings for the test. The path is as follows: Start > Programs > Instron > Merlin > Methods > Compressive > PM_zinc_ME352.

8. The force limits in this method are based on the following calculations for 150 MPa and 450 MPa pressure:

$$Pressure = \frac{P}{A}$$

   P (in lb) is the force limit in the software method
   A is the cross-sectional area $(\pi/4)D^2$, where D (diamater) = ½ in

Making sure that pressure is converted to English units (since the machine is set to those units), we get:
   P_1 (first load) = 4271 lb (for 150 MPa)
   P_2 (second load) = 12812 lb (for 450 MPa)

9. The force limits need to be changed in the software method for testing at the two different pressures. Once the method is opened, click Test > Limits.

10. Run the test by going to the home screen in the Merlin software and clicking the "Run Test" button. Specify the output data path when prompted and click "Start Test" to begin.

11. When the force limit is reached, compression will stop. Manually jog the stage down (using the Instron controls) and remove the test sample carefully. At the low pressure test, the zinc tablet (compacted powder) will be fragile.

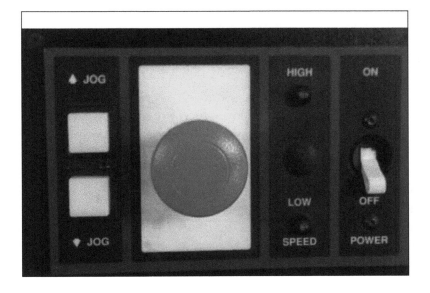

Figure 5. A control panel used in this compression experiment. It controls start/stop of the machine, as well as the jogging of grips up and down.

12. The pressed face of the tablet will have a lip. Carefully file it down till it disappears. Do not file beyond the lip base.

13. Transport the sample to the microhardness tester. Make sure you remember which side was pressed (the side facing up).

Vickers hardness test and microscopy:

1. Mount the sample in the microhardness tester with the compressed face up.

2. Perform the Vickers hardness test. Make sure the indentation is close to the center of the sample, as compaction density is not perfectly uniform throughout the surface area.

3. Each group member will record the diagonals of the indent using the high magnification scope (at 40X). The diamond indents will be much more difficult to see in your samples compared to the finely polished sample we used in the hardness testing lab.

4. Remove the sample from the microhardness tester and take it to the light microscope.

5. Mount the sample in the microscope and find the indent using the software linked to the microscope.

6. Save the image to a USB drive and include it in your lab report.

7. Record the thickness of each tablet using the digital caliper (0.0005 in accuracy.)

TO-DO LIST

1. Calculate the Vickers hardness for each sample.

2. Calculate the mass density for each sample. (mass/volume).

3. Compare your results for the high pressure sample to the low pressure sample in terms of hardness and density. Explain if the difference makes sense to you and why.

4. Compare your density results with tabulated values for pure zinc.

5. Why was it important that we tested the compressed (top) face of the tablet instead of the bottom face?

6. Discuss Vickers hardness measurements taken by each group member. Were the values close to each other or far apart? Explain why.

7. How was the Vickers test for your samples different from the provided calibration sample used in the hardness lab?

Note to Instructors: You can augment or edit the above procedure or To-Do List in any way that is most suitable for your lab/class situation. For example, the digital caliper (for measuring tablet dimensions) can be replaced with an analog one. In this process, the alignment of the plunger with the cavity in the cylindrical die is essential for not damaging the hardware during compression. More than two compaction loads can also be used in this experiment. Also, make sure that your compaction load does not exceed the machine capacity (including load cell capacity). The cylindrical tablet diameter is dependent on the cavity diameter in the cylindrical die. Also, the weight of the tablet is dependent on how much powder was put into the mold before pressing. Both of these parameters (diameter and weight) can be controlled by the user.

CHAPTER 8
STRESS-STRAIN BEHAVIOR
(EXPERIMENT/LAB 6)

To study the relationship between stress and strain, ductile and brittle material specimens of cylindrical geometry are loaded in uniaxial tension here. The applied load and elongation are measured throughout the test up to fracture/failure. Observations of the elastic, yield, necking, and fracture/failure behavior, as well as a comparison of ductile and brittle behavior are made.

BACKGROUND

Review the sections of your Materials Science book dealing with the uniaxial tension test, strength, ductility, toughness, and true stress and true strain.

INTRODUCTION

In the engineering design of structures, components, and machines, the "elastic" properties of the materials used must be accurately known. Failure is the ultimate condition to be avoided in service, but the definition of failure depends on the application. In many cases, yielding or the onset of plastic deformation is defined as failure. Often, however, necking constitutes failure, and in virtually all cases, a fractured/ruptured part is considered to have failed. The nature of structural failure may be studied experimentally, but only through "destructive testing." A machine part may fail in tension, compression, torsion, bending, fatigue, etc. Destructive tests may be applied to materials to isolate each kind of failure. Such testing is satisfactory if a "sample" of the material is used. If, however, the structure, machine, or part is to be checked for defects or for conformity with specifications, some form of nondestructive testing must be employed.

Structural materials such as steel, aluminum, or plastics are complex substances. Many of the properties of these materials vary somewhat with small differences in composition or processing treatments. Hence, quantitative values must be determined specifically for every sample through experimental testing.

The most common material property information, needed by the designer, is contained in each material's stress-strain curve. Most metals and many other structural materials must be used chiefly within their linear range, where the strain or stretch is directly proportional to the applied stress or force, respectively. Actual stress-strain curves for typical materials show that elastic deformation may be only a small part of the total data set.

The two related quantities of interest in a simple tension test are stress and strain. Stress is just a normalized load and is defined as the force or load per unit area. In a uniaxial tension test, it is assumed that the force distribution is uniform on any plane inclined to the axis of loading. The normal stress component on a plane perpendicular to the loading axis is defined as:

$$\sigma = \frac{P}{A_o} \tag{1}$$

where P is the applied load, and A_o the initial cross-sectional area (i.e., the cross-sectional area before loading). The strain, ε, is the elongation per unit length of the cylindrical bar. If the deformation is uniform along the length of the rod, then the strain component along the axis is defined as:

$$\varepsilon = \frac{\Delta l}{l_o} \tag{2}$$

where Δl is the change in length or elongation and l_o is the original length of the rod over which elongation is measured (i.e., the gauge length). $\Delta l = l - l_o$, where l is the current or instantaneous gauge length corresponding to the current force P. The results of a standard tension test are represented in the form of a stress-strain diagram. Specifically, σ and ε defined here are called the engineering (or nominal) stress and the engineering (or nominal) strain.

Linear-Elastic Properties:

The behavior of materials in simple tension may be divided into elastic and inelastic (or plastic) ranges. Elastic action means that the material returns to its initial state after the removal of a load; it is a reversible process. For most engineering materials, such as steel, the relationship between stress and strain for very small strains is, to a close approximation, linear (i.e., Hooke's law is followed). This type of material is generally called linear-elastic or just elastic, and a typical stress-strain curve is shown in Figure 1. The stiffness of a linear-elastic material under simple tension is measured by the Young's modulus E, which is the slope of the stress-strain line:

$$E = \frac{\sigma}{\varepsilon} \tag{3}$$

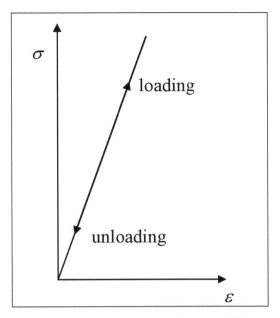

Figure 1. Schematic illustration of the stress-strain curve of a typical linear-elastic material.

The **elastic limit** is the highest stress at which the behavior of the material remains elastic, i.e., suffers no permanent deformation. The **proportional limit**, is the highest stress at which Hooke's law is valid (symbolized by σ_{pl}). In many engineering materials Hooke's law holds throughout the entire elastic range and in such cases, the elastic limit and proportional limit coincide.

The **elastic resilience** is the capacity of the material to absorb energy in the elastic region. The elastic energy per unit volume of the rod (corresponding to a current strain value of ε_1) is given by:

$$u_e = \int_0^{\varepsilon_1} \sigma d\varepsilon \tag{4}$$

where σ and ε are defined in equations (1) and (2). Equation (4) represents the area under the stress-strain curve from zero strain to strain ε_1. The **modulus of resilience**, u_r, is defined as the elastic strain energy stored per unit volume when loaded up to σ_{pl}, the proportional limit. From equations (3) and (4), it may be shown that:

$$u_r = \frac{\sigma_{pl}^2}{2E} \tag{5}$$

Inelastic or Plastic Range:

As the tensile force applied to the specimen increases, a point is reached at which irreversible micro-structural changes take place in the material. After this point, the behavior is said to be inelastic or plastic (at least in part). This point is called the **yield point** and the stress value corresponding to it is termed the **yield strength** or **yield stress** of the material (denoted as σ_y or σ_{YS}). Correspondingly the strain value at this point is called the yield strain (denoted by ε_y).

The total strain after this point may be divided into two components: the irreversible part that is the plastic or permanent strain and the recoverable part that is the elastic strain. These two components are illustrated in Figure 2. Notice that the stress-strain relationship after this point becomes curved or non-linear (i.e., a departure from the early-on linear elastic behavior).

An important property of the plastic strain is that it is not accompanied by any significant change in volume. Elastic strain under uniaxial tension, however, produces normalized increase in volume equal to $\varepsilon(1-2v)$ in simple tension where v is Poisson's ratio. Poisson's ratio v is a material property defined in the elastic region of the stress-strain behavior as the negative of the ratio of lateral strain to longitudinal strain. Longitudinal strain is the axial strain along the tensile axis and lateral strain is the strain along a normal direction to this axis.

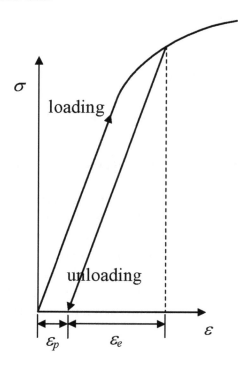

Figure 2. The elastic and plastic components of strain in the plastic region of a stress-strain diagram [ε = total strain = $\varepsilon_e + \varepsilon_p$ (6)] During unloading, only the elastic component is reversed. Thus, the unloading curve is linear and has slope E. The strain remaining after unloading is the permanent or plastic strain.

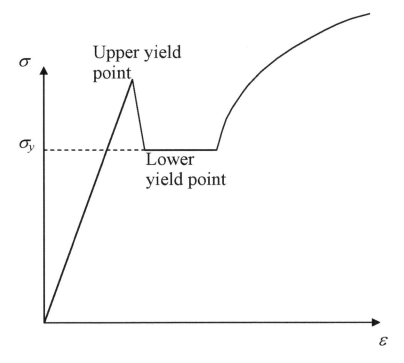

Figure 3. Schematic stress-strain curve for a material with a sharp yield point. The upper yield point is the highest stress prior to yielding, and the lower yield point is the lower stress just after yielding.

For some steels and other materials, yielding occurs very suddenly, resulting in large plastic deformation without any corresponding increase in stress or, perhaps, even a lowering of stress. Such upper and lower yield points are illustrated in Figure 3. Where there are two yield points, the lower yield point stress value is the value conservatively used in engineering analysis.

Usually, however, the transition from elastic to plastic behavior is more gradual. In these cases, the definition of a precise yield point is more difficult (since the elastic limit is beyond the proportional limit) and an offset yield stress definition is used. This yield stress is defined by drawing a line (from the strain axis) with slope E from a strain value typically taken as 0.002 (0.2%); see Figure 4. Where this line intersects the stress-strain curve is defined as the yield point, with its corresponding stress value defined as σ_y or σ_{YS}.

Generally, materials work-harden after yielding, meaning that the stress required for continued deformation increases as the plastic strain increases. As the elongation proceeds with increasing load, a point is reached where the specimen cannot sustain any higher load (this point on an engineering stress-strain diagram is termed the Ultimate Tensile Strength point or simply the Tensile Strength point); see Figure 5. The stress value corresponding to this peak or highest point on the stress-strain diagram is termed the **Ultimate Tensile Strength (UTS)** or simply the **Tensile Strength (TS)** of the material. For a ductile material, deformation in the bar after this point ceases to be uniform along its length (i.e., this point signifies the termination of uniform plastic flow and the commencement of "necking" behavior in the bar/rod). More on necking later.

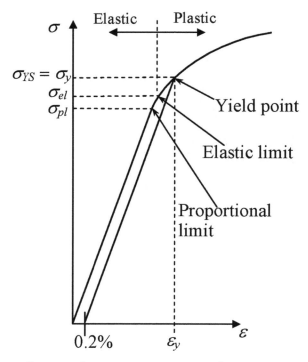

Figure 4. Schematic illustration of a typical stress-strain curve in the region around yielding showing the definition of the 0.2% offset yield stress.

Although the strength of a ductile metal increases as it is plastically deformed, its cross-section becomes smaller. Thus, the actual stress on the material, defined as load divided by the cross-sectional area, is increasing during plastic deformation. So long as the increase in *strength* due to work-hardening is greater than the increase in *stress* due to the decreasing area carrying the load, plastic deformation is stable. Work-hardening, however, is due to microstructural processes which cannot continue indefinitely and gradually ceases. Eventually, there comes a point where the increase in *strength* due to work-hardening is not sufficient to compensate for the increase in *stress*. At this point, deformation concentrates in the region where this first occurs. Thus, further plastic deformation is localized to a narrow region which is called a "neck." The load required to cause further deformation after this point drops steadily with further plastic strain and with reduction in neck area. The deformation goes on until the specimen ruptures (see Figure 5). The necking behavior described above is called a plastic instability because once it occurs, failure is inevitable. This type of necking is common in ductile metals.

The failure of a ductile material is commonly referred to as "rupture" as opposed to "fracture." On the other hand, failure of brittle materials is commonly referred to as fracture. For a brittle material, the inelastic or plastic region (compared to the elastic region) is not lengthy compared to ductile materials (i.e., it has a short strain range). Therefore, for brittle materials, fracture happens soon after yielding, to the extent that a yield point is not typically defined for these materials. Therefore, no necking behavior is observed with true brittle materials.

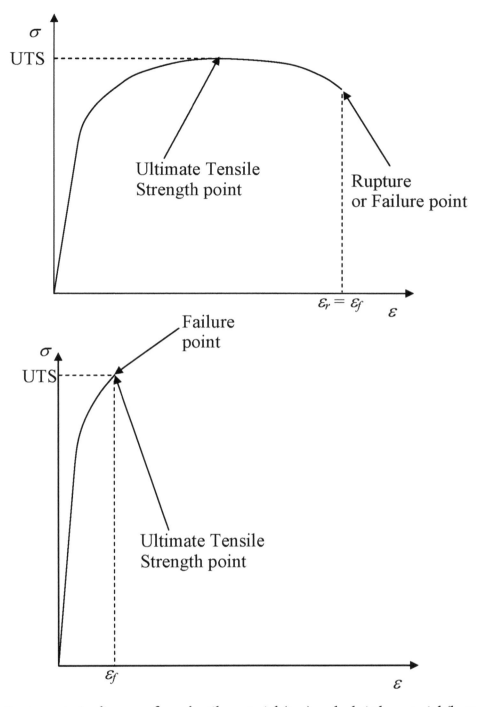

Figure 5. Schematic stress-strain diagrams for a ductile material (top) and a brittle material (bottom). Brittle materials fail by cracking soon after the linear-elastic region, while ductile materials work-harden after yielding, neck and rupture beyond the UTS point. Here, ε_f is the "failure strain" (also called the "strain-to-failure").

Two additional quantities usually obtained from a standard tension test are the **percent elongation** and **percent reduction in area** at rupture. Both of these quantities are measures of the ductility of a material (i.e., the amount it deforms prior to failure) and are defined as follows:

$$\textbf{percent elongation} = \frac{l_f - l_o}{l_o} \times 100 \qquad (7)$$

$$\textbf{percent reduction in area} = \frac{A_o - A_f}{A_o} \times 100 \qquad (8)$$

where l_f and A_f are the gauge length at failure (i.e., final gauge length) and the cross-sectional area of the failed neck (i.e., final neck cross-sectional area). Also, l_o and A_o are the initial (before loading) gauge length and the initial cross-sectional area of the tensile specimen, respectively. The **uniform elongation**, which is the engineering strain at the start of necking (i.e., the strain at the UTS), is also sometimes used as a measure of ductility in ductile metals.

Toughness is the overall capacity of a material to absorb energy, both elastic and plastic energy, prior to fracture. The **modulus of toughness** is defined as the work done on the material per unit volume during the tension test and is given by the integral:

$$u = \int_0^{\varepsilon_r} \sigma d\varepsilon \qquad (9)$$

which is integrated from zero strain up until rupture. This integral is equal to the <u>total area under the stress-strain curve</u>.

True-stress, True strain, and Strain-hardening:

For strains small compared to unity, no ambiguity results from the use of the definitions given in equations (1) and (2). The cross-sectional area and the gauge length remain practically unchanged from their initial values (e.g., the percent change in length for cold-worked 304 stainless steel is about 0.27% for a stress value of 515 MPa!).

Later in the tension test, both the reduction in area and the strain become very large for ductile materials. In other words, neither A nor l is constant, and the stress and strain as defined in equations (1) and (2) are no longer accurate for many analysis purposes. It is often desirable to represent our tension test data on the so-called true stress-true strain diagram. **True stress** σ_t is defined as:

$$\sigma_t = \frac{P}{A} \qquad (10)$$

where A is the current or instantaneous cross-sectional area of the specimen corresponding to load P. An increment of **true strain**, $d\varepsilon_t$, is defined as the ratio of the change in length, dl, to the instantaneous length l:

$$d\varepsilon_t = \frac{dl}{l} \qquad (11)$$

In a tension test, if the initial length is l_o, then when its length has increased to l_1, its natural or true strain is defined as a summation (i.e., integral) process as follows:

$$\varepsilon_t = \int_{l_o}^{l_1} \frac{dl}{l} = \ln\left(\frac{l_1}{l_o}\right) \tag{12}$$

Stress and strain as defined in equations (1) and (2) are called nominal, or engineering stress and strain, as distinguished from the true stress and strain in equations (10) and (12).

Figure 6 gives a comparison of stress-strain diagrams for a ductile metal using nominal and true values. As may be expected, the two diagrams practically coincide over the elastic range and the beginning of the plastic range.

Figure 6. Schematic of the nominal and true stress-strain curves for a ductile metal.

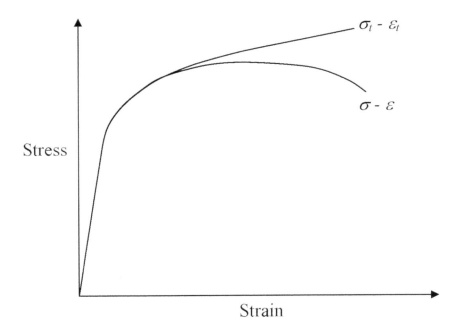

It should be noted that the true stress does not fall off after the engineering UTS point is reached, showing that the characteristic decrease in stress in the nominal stress-strain curve is caused by a change in geometry of the specimen and is not inherent in the material. The material continues to work-harden, but at a rate that is insufficient to overcome the increase in applied stress due to the decrease in cross-sectional area. Because the cross-section continues to decrease in the necked region, the load required for further deformation also decreases.

As mentioned previously, when the plastic strain is large compared to the elastic strain, it may be assumed that the deformation proceeds at constant volume:

$$A_o l_o = Al \tag{13}$$

This assumption is used in computing the true stress from the engineering stress-strain data. It may be shown that true stress can be expressed as:

$$\sigma_t = \sigma(1+\varepsilon) \tag{14}$$

provided that the assumption of constant volume is valid. True strain can be expressed as:

$$\varepsilon_t = \ln(1+\varepsilon) \tag{15}$$

After necking, deformation is localized to the neck region, and equation (13) is <u>no longer valid</u>. Therefore, equations (14) and (15) cannot be used to calculate the true stress and true strain beyond the UTS point. In order to calculate the true stresses and true strains beyond the uniform plastic range, the cross-sectional area of the neck region must be known as a function of the load. Since this information is generally not taken during a typical tension test, the true stress-true strain curve is extrapolated from the point corresponding to necking to the values at failure given by:

$$(\sigma_t)_f = \frac{P_f}{A_f} \tag{16}$$

and

$$(\varepsilon_t)_f = \ln\left(\frac{A_o}{A_f}\right) \tag{17}$$

where P_f is the load at failure and A_f is the final cross-sectional area of the neck measured from the broken specimen.

For analysis in the plastic range, it is desirable to find an equation that approximates the stress-strain curve in this range. It has been found that for many, but not all, metals the empirical equation:

$$\sigma_t = k\,\varepsilon_t^n \tag{18}$$

is a good approximation. The constants, k and n, depend on the material and are usually called the **strength coefficient** and the **strain-hardening exponent**, respectively. Taking the natural logarithm of equation (18) one obtains:

$$\ln\sigma_t = \ln k + n\ln\varepsilon_t \tag{19}$$

The constants, k and n, can therefore be found from a ln-ln plot of true stress and true strain. The intercept of such a plot at $\ln\varepsilon_t = 0$ (or $\varepsilon_t = 1$) is $\ln k$ and the slope of the resulting line is n. If the data plotted on a $\ln\sigma_t$ - $\ln\varepsilon_t$ plot is not linear, then equation (18) does not hold for the material tested.

Apparatus

The device used for applying and measuring the load in this lab is called an Instron Tension/Compression Testing Machine (see Figure 7). There are other types of tension testing machines that differ primarily in how the crosshead is moved to strain the specimen. Instruments for measuring the length of the test section during a test will depend upon the material and the purpose of the test. A strain gauge can be mounted directly to the specimen. This is the most accurate method, but it is also the most expensive and time-consuming, since the strain gauge must be mounted to the specimen and cannot be removed and reused once it is. Strain gauges mounted directly to the specimen are often used when accurate measurement of the elastic properties is required. At large strains, however, strain gauges will often become separated from the specimen (i.e., peels off from the specimen surface). For routine strength measurements and tests requiring plastic deformation of the specimen, a mechanical extensometer is most commonly used. This device is attached to the gauge points at each end of the test section, and measures the average strain over the test section's gauge length. Extensometers are V-shaped electromechanical devices and are calibrated to give the change in separation of the gauge points. This is done through measuring voltage changes that are converted to displacements/extension. Extensometers are generally accurate enough for most engineering applications and have the advantage that they may be used more than once.

Figure 7. The Instron Universal Tension/Compression Testing machine used in the experiment.

Tension Test Specimens may be round or flat. They have a reduced test section in the center, and ends suitable to gripping in the particular testing machine set-up being used. The ends may be plain, threaded, shouldered, or pin-connected. The gauge length, over which the dimensional extension (strain) measurements are made, should be at least four times the minimum diameter. For metals, the most common gauge length is 2.00 inches with a diameter of 0.50 inches.

In the lab, you will be given four specimens: steel, gray 20 cast iron, aluminum, and brass/copper.

Figure 8. Schematic showing the dimensions of a common tension specimen with a circular cross-section.

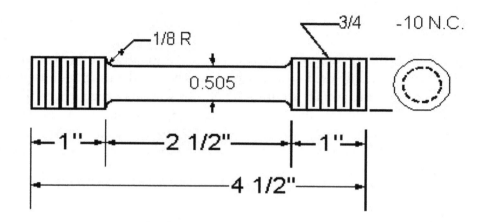

Procedure

Because the initial dimensions vary from specimen to specimen and are needed to calculate the engineering stress and strain, the mean diameter and the gauge length of each specimen must be measured prior to testing. Measure the diameter (or width and thickness) of each specimen four times, having a different student in the group perform the measurement each time. If a measurement is obviously wrong, repeat it until the variation between measurements is small. Measure and mark two inches about the center of the gauge section of each specimen.

	Steel	Cast Iron	Aluminum	Brass/Copper
Original Diameter, in				
Original Area, in^2				
Original Gauge Length, in				
Final Gauge Length, in				
Failure Diameter, in				
Area at Failure, in^2				

1. Turn on the machine.
2. All the digital display screens on the control panel must show [-1.8.8.8.8].

3. Then this number disappears and the upper three digital displays turn blank (showing no numbers) and the lower one must show [3399], which later on disappears as well.

4. Mount on your cylindrical sample.

5. Using the control panel, press Cal (calibrate) Load and Enter then Bal (balance) Load and Enter, then press Cal Strain and Enter.
6. Mount the Extensometer and try to place it at the middle of your sample using rubber bands.

7. **Remove the Extensometer pin.**
8. Plug in the IEEE cable.
9. Press Bal Strain and then the Enter button.

10. Set the crosshead speed at 0.1 in/minute.
11. Run the experiment using (Merlin) a computer program that is connected to the Instron machine.
12. While running the experiment the program will prompt you to remove the Extensometer to resume running the test.
13. A tensile load will be increasingly applied on our sample until it fails and breaks up. The data will be stored in a file. It is encouraged that all data analysis be performed using MS Excel.

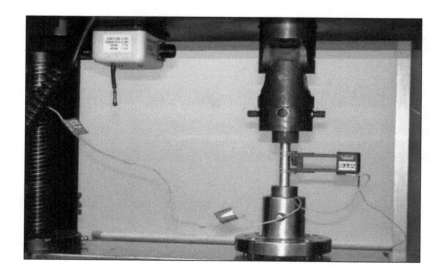

TO-DO LIST

1. The test data must be turned in for the tension test data analysis. Clearly show all calculations as if it were a homework problem. If Microsoft Excel or some other spreadsheet is used to perform calculations, then include a printout of the spreadsheet and clearly write which equations were used in each column where calculations were made. Also, include the raw data from the experiment initialed by the Teaching Assistant (TA) during the laboratory period. The specimen dimensions and other recorded data must be neatly handwritten and initialed by the team members prior to leaving the lab.

2. Plot the engineering stress-engineering strain curve. This curve must pass through the origin of your plot. If it does not, make the necessary corrections.

3. To determine the Young's modulus, plot the engineering stress-strain curve only to strains just past the yield point of the specimen. By plotting only the elastic region, the strain axis is expanded, allowing a more accurate determination of Young's modulus, proportional limit, and 0.2% offset yield strength. The proportional limit, 0.2% offset yield point, ultimate tensile strength, and rupture stress shall all be clearly labeled on this graph and the graph in part (2), as appropriate. If the labels cannot be produced on your graph using a computer, then label them neatly in ink. Do not photocopy these graphs.

4. Plot the true stress-strain curve in the same graph as (2). Clearly show the point at which necking occurs, and use a dashed line to show the extrapolated curve beyond that point.

5. Derive equation (17).

6. Obtain the ln-ln plot of the true stress- true strain diagram over the plastic range. Determine the range over which equation (18) is valid, and use equation (19) to find k and n.
   Putting the results for all materials tested in the same table, tabulate the following:
   a. Young's modulus
   b. Proportional limit
   c. σ_{YS} using a 0.2% offset
   d. Modulus of resilience
   e. UTS and uniform elongation
   f. Percent elongation
   g. Percent reduction in area

    h. True strain at rupture/failure

    i. Engineering stress at rupture/failure

    j. True stress at rupture/failure

    k. Modulus of toughness

    l. Strength coefficient

    m. Strain hardening exponent

7. Discuss the specimen's fracture surface shapes in light of its ductility. Did your specimens fracture/fail in the expected manner? Descriptions shall include material composition and fracture appearance (i.e., flat, granular, cup-cone, shear, irregular, etc.).

Note to Instructors: You can augment or edit the above procedure or To-Do List in any way that is most suitable for your lab/class situation. One of the important things about the above procedure is that it be performed at a low strain rate. In fact, a typical stress-strain diagram is established at very low strain rates. If the strain rate is much higher, the resulting stress-strain diagram will differ as the mechanical behavior of most materials is strain-rate sensitive. Note that no matter what experimental setup is used to perform the above experiment, syncronized elongation-force data pairs must be recorded for establishment of an engineering stress-strain diagram.

CHAPTER 9
EFFECT OF WORK HARDENING ON MECHANICAL PROPERTIES (EXPERIMENT/LAB 7)

One objective here is to learn what "strain-hardening" or "work-hardening" of a material means. Another objective is to determine the effect that strain-hardening, or work-hardening, has on the hardness measurements of a metal.

<u>BACKGROUND</u>

Consider the following typical stress-strain diagram (typical for ductile metals):

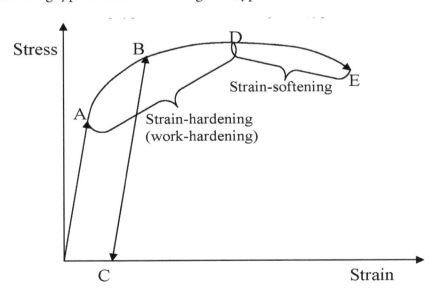

Figure 1. A schematic representing an engineering stress-strain diagram typical of a ductile metal.

The initial yield point for the material was at point A. IF the material or specimen is unloaded at point B down to zero load at point C and then RE-LOADED back, the new yield point for this specimen will be at point B (notice that the line between points B and C is parallel to the original line ending at point A, i.e., they both have the same slope or Young's modulus/modulus of elasticity). In both cases, the ultimate tensile strength (UTS) or simply tensile strength (TS) of the materials is at point D. Point E represents the failure/rupture point where the rupture/failure strain occurred.

What is noticed from this diagram is that the re-loaded specimen (point C to points B, D, and beyond) is essentially a new specimen with a higher yield point (point B compared to point A). In other words, this material got stronger because of the initial plastic straining it experienced in its history. The above-mentioned behavior is valid for any point between point A and point D. We call the region between points A and D a "strain-hardening" or "work-hardening" region of the curve. Hardening is synonymous with strengthening in this context here. As is learned in introductory materials science textbooks, there is a relationship between the strength of a material and its hardness, and hence the synonymy.

What causes strain-hardening in metals like aluminum or steel alloys is the fact that after the initial yield point A, a lot of dislocations inside the metal have generated to account for the plastic strain encountered by the material. The more the material is strained beyond point A, the more dislocations are generated and the higher their density. The higher the dislocation density, the more encounters between dislocations. Many of these encounters cause dislocation pinning and hindrance of the dislocations/further plasticity. So it requires more stress beyond point A to keep flowing the material plastically in order for the dislocations to overcome the generated pinning points and obstacles.

Indeed, it was noted in another chapter that the strain-softening behavior observed beyond the UTS (or point D in Figure 1) in an engineering stress-strain diagram is due to geometric changes in the specimen (in particular necking in the specimen) and not due to inherent material softening. In fact the material continues to harden using the microstructural mechanism mentioned above. This fact is observed by plotting the true stress – true strain diagram instead of the engineering or nominal one. Figure 2 shows the true stress – true strain diagram overlaid on top of the engineering one. The figure shows continued strain-hardening (i.e., work-hardening) in the true stress-strain diagram.

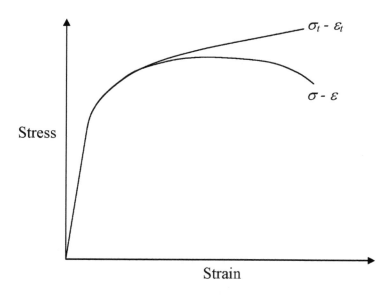

Figure 2. Schematic of the engineering (nominal) and true stress-strain curves for a ductile metal.

Procedure

1. Load Steel 2018 HR (hot rolled) tensile test specimens under a tensile load. Load each specimen until it breaks at point E (Figure 1). A broken cylindrical specimen will look something like Figure 3 (a typical cup-and-cone fracture surface for aluminum).

2. Cut the tensile-tested specimens into cylindrical pieces between the dashed lines (about 1 cm each in length/height). You might generate 3 or 4 such pieces.

3. If the cut surface is relatively flat, then move to the next step. If it is not, then flatten it with careful grinding.

4. Test each piece in a Rockwell tester (a microhardness tester requires polishing of samples). Target scale B for aluminum and many steels.

5. For accurate hardness readings, make sure you stay away from the edges of a piece, as well as from other pre-existing indentations. Indent in a new spot about 4–5 indentation diameters away from edges or old indents.

6. Record all hardness numbers.

7. Also test a pristine, or pre-worked, specimen. Keep this hardness number to compare with the hardened material.

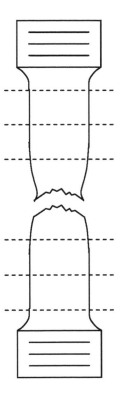

Figure 3. A metallic tensile specimen broken into two pieces. Each piece is then sectioned (shown by the dotted lines) in parts along the specimen length away from its two ends.

1. Define what strain-hardening or work-hardening is.
2. Explain why a strain-hardened material is harder than the original pre-worked material.
3. Get the average hardness and standard deviation of hardness measurements, based on all the hardness readouts collected.
4. How much harder, on average, was the hardened material compared to the original pre-worked specimens? Do this on a bar graph showing the span of the readings as well as the averages.

Note to Instructors: You can augment or edit the above procedure or To-Do List in any way that is most suitable for your lab/class situation. It is important to keep the strain-rate constant in the different tensile specimens broken for hardness measurements. Keep in mind that the strain-rate could affect the mechanical properties. If one elects to do microhardness testing instead of Rockwell, it is important to use the largest indentation load possible and be consistent with this load across the different samples indented. Lastly, do not indent the pieces (or parts) of the broken specimen that are inside the neck region as these might have much different hardness values from the rest of the specimen length.

CHAPTER 10
PAPER TOWEL TENSION
TEST (EXPERIMENT/LAB 8)

Students will perform a tensile test on a paper towel. Statistical analysis and knowledge of experimental error will be applied to further understand the results of this experiment/lab.

INTRODUCTION

Manufacturers of paper have to meet strict requirements for the strength, thickness, weight, smoothness, and appearance of their product. They have to do this in order to consistently meet the needs of their customers and get repeat business. There are also other considerations imposed on them by standards. For example, bathroom tissue paper cannot exceed a certain strength, otherwise it won't disintegrate when flushed with water, and that would cause clogging of the plumbing network.

In this lab, students will mount a rectangular piece of towel paper in a tensile testing machine and apply tensile force until the paper breaks or ruptures. This will be done for at least one of the principal directions of the paper (Machine direction and/or Cross-machine direction).

Tensile strength, tensile energy absorption, and stretch will be recorded for each sample.

- Tensile strength is indicative of the strength derived from factors such as fiber strength, fiber length, and bonding, as well as strength produced by the addition of wet and dry strength resins.
- Stretch is indicative of the ability of paper to survive non-uniform tensile stress.
- Tensile energy absorption is a measure of the ability of a paper to absorb energy (at the strain rate of the test instrument), and expresses the "toughness" of the sheet.

Definitions

Tensile Strength—The maximum tensile force developed in a test specimen before rupture on a tensile test carried to rupture under prescribed conditions. Tensile strength is the force per unit width of the test specimen.

Stretch—The maximum tensile strain developed in the test specimen before rupture in a tensile test carried to rupture under prescribed conditions. The stretch (or percentage elongation) is expressed as a percentage, i.e., one hundred times the ratio of the increase in length of the test specimen to the original test span.

Tensile energy absorption (TEA)—The work done when a specimen is stressed to rupture in tension under prescribed conditions, as measured by the integral of the tensile strength over the range of tensile strain from zero to maximum strain. The TEA is expressed as energy per unit area (test span × width) of the test specimen.

Procedure

1. Latex gloves must be worn when handling paper. Any moisture transferred to the paper towel will greatly affect the results of this experiment.
2. Cut paper into 3 in × 5 in rectangles. See Figure 1. Each group will need four samples.

Figure 1. A paper towel specimen cut into a 3 in × 5 in rectangle.

3. Clamp the paper towel with the jaws on the tensile testing machine. Make sure there is no slack and the paper is secured with an appropriate clamping pressure where there is no slippage, but that the paper is not damaged by the clamping force. Use an appropriate clamping pressure. Also make sure that the sample is free from any abnormalities such as creases, wrinkles, or perforations.
4. Set the clamps to an initial test span (nominal value is 3 in). See specimen under stretching force in Figure 2.
5. Set the crosshead separation speed to 2 in/min.
6. Carefully remove slack in between the jaws, but not to the point where the paper is strained.
7. Test 10 samples in each principal direction.
8. Reject any sample that slips in the clamps or breaks at the jaws, or any sample that breaks within 5mm of the clamp area (if further inspection indicates the break location is due to improper clamping conditions or misalignment).
9. Record the tensile strength to the nearest 0.1 N/m (1 gf/3 in), the stretch to 0.1% and the tensile energy absorption to 0.01 J/m^2 (0.0007 ft-lb/ft^2).

TO DO LIST

1. Report properties for each test sample with a resolution of three significant digits.
2. Find the average, standard deviation, and range for tensile strength.
3. Find the average, standard deviation, and range for stretch as a percentage.
4. Find the average, standard deviation, and range for the tensile energy absorption.
5. Report the number of test samples that were rejected and the reasons for rejection.

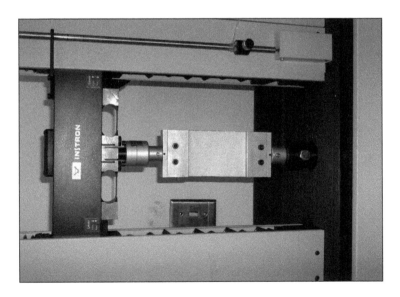

Figure 2. A paper towel specimen under tension between the two jaws of an Instron tabletop tension/ compression machine. Each jaw can clamp onto 1–1.5 in of paper.

Note to Instructors: You can augment or edit the above procedure or To-Do List in any way that is most suitable for your lab/class situation. One important thing when cutting the sample papers is to use a sharp scissor or cutting blade such that no rough or broken edges are left in the specimens after cutting them. Also, make sure that cut sides are parallel to each other within 1°. Lastly, make sure the two grips are in the same plane and their edges parallel to each other within 1°. Note that unless you buy expensive jaws for this testing procedure, it is not too difficult to fabricate them in-house. This procedure is also applicable to other types of papers besides paper towel (i.e., facial and bath tissue, as well as napkins).

CHAPTER 11
MATERIAL TOUGHNESS MEASUREMENTS (EXPERIMENT/LAB 9)

To investigate the fracture behavior of different materials, the Charpy Impact Test is used. The fracture toughness of a BCC steel material is compared to that of an FCC alloy, and the difference in fracture appearance and impact toughness of ductile and brittle materials is also investigated.

BACKGROUND

Review the chapters in your materials science book dealing with ductile and brittle fracture and impact testing/impact toughness or impact energy.

INTRODUCTION

Fracture or breakage of a material may be defined as the mechanical separation of a solid owing to the application of stress. Fractures of engineering materials are broadly categorized as ductile or brittle, and fracture toughness is related to the amount of energy required to create fracture surfaces. In ideally brittle materials such as glass, the energy required for fracture is simply the intrinsic surface energy of the material. For structural alloys at room temperature, considerably more energy is required for fracture because plastic deformation accompanies the fracture process.

Unlike other testing applications that we have studied, impact testing involves the sudden and dynamic application of the load. Parts such as shafts, bolts, anvils, and dies are examples of items subjected to impact loading. **Impact Toughness** is defined as the resistance of a material to rapidly applied loads. Impact Toughness is a material property, which is the capacity of a material to resist fracture/breakage/rupture when subjected to impact.

The toughness of a material is a measure of its ability to absorb energy prior to fracture. Ductile fracture is preceded by extensive plastic deformation. Under a microscope, ductile dimples are often observed, and at the edges of the fracture surface near the specimen surface, shear lips are usually seen without the aid of a microscope. Because plastic deformation requires energy, ductile materials tend to have relatively high fracture toughness.

Brittle fracture occurs by separation of the material at grain boundaries (intergranular fracture) or through grains (transgranular fracture). Brittle crack propagation is very rapid. It is accompanied by little or no plastic deformation and thus requires very little energy. For this reason, brittle materials generally have low fracture toughness.

There are several ways of measuring the toughness of a material. As seen in another chapter, an ordinary uniaxial tension test provides a measure of toughness. In such a test, the energy absorbed in breaking the specimen is equal to the area under the stress-strain curve. In uniaxial tension testing, however, the strain rate is generally low, typically 10^{-4} - 10^{-2} s^{-1}. To examine the fracture behavior of materials under more rapid loading conditions, an impact test is often used. Impact testing typically involves swinging a pendulum hammer through the specimen to break it. The difference between the initial and final heights of the pendulum is related to the energy absorbed in fracture or rupture.

Impact Toughness generally refers to the toughness of a material when loaded dynamically with an impact force and in the presence of a notch or crack (which acts as a stress concentrator or stress riser). For this, notched specimens are usually used. Various types of notched-bar impact tests are used, but two types of tests are most common:

Charpy V-notch Impact Test (Figure 1)
Izod V-notch Impact (Figure 2)

Specimens for both tests are the same: prismatic with a square cross-section and 45° notches at a depth of 2 mm. In a Charpy impact test, the specimen is supported as a beam in a horizontal position with the V-notch facing away from the incoming impact. In an Izod impact test, the specimen is vertically clamped with the V-notch just exposed above the clamp's surface and facing into the incoming impact. The impact of a heavy swinging pendulum forces the specimen to bend and fracture at a very high strain rate (on the order of $10^3 s^{-1}$).

The data obtained from an impact test corresponds to the energy expended in fracturing the specimen. After breaking the test bar, the pendulum swings through to a height that is decreased from the original height by a value corresponding to the amount of energy that was absorbed in breaking the specimen. This energy, measured in ft-lb or J, can be directly read from a calibrated dial on the machine and is a function of temperature. The fracture energy measured in an impact test is only a qualitative measure of the fracture toughness. Other more quantitative measures are used for design purposes. Charpy impact testing is useful, however, for qualitative comparisons of different materials at the same temperature or of the same material at different operating temperatures.

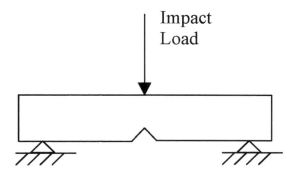

Figure 1. Schematic illustration of loading in a Charpy Impact Test.

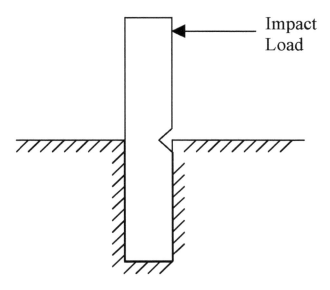

Figure 2. Schematic illustration of loading in an Izod Test.

Other information that may be obtained from the results of an impact test comes from an examination of the fracture surface. A fracture surface may be described as ductile, brittle, or a mixture of the two. A brittle fracture surface, characterized by flat facets, provides high reflectivity and a bright appearance. A ductile fracture with a dimpled, fibrous character, on the other hand, provides a light-absorptive surface and a dull appearance. An estimate should always be made as to what percentage of the fracture surface may be deemed brittle, although this is best done using an optical (light microscope) or a scanning electron microscope (SEM).

In summary, impact tests can be used to determine:
- Impact Energy absorbed for each specimen, in ft-lb or J
- The nature of fracture: Ductile or Brittle
- Percentage of brittle fracture on the fracture surface

Equipment

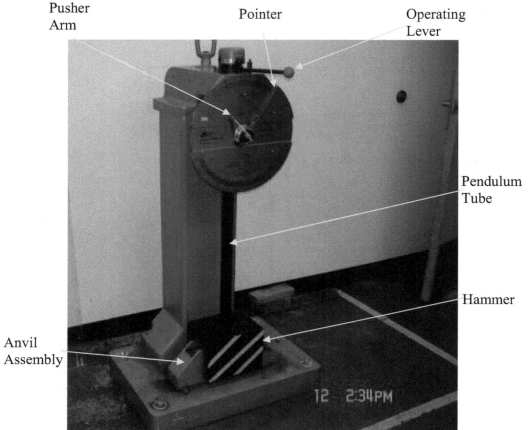

Figure 3. The Baldwin SI-1 Charpy impact testing machine.

Figure 3. The Baldwin SI-1 Charpy impact testing machine.

A pendulum-type impact testing machine, the Baldwin SI-1 can be adapted easily to Charpy or Izod impact testing. It measures the energy required to break a specimen by measuring the difference between the angle of rise when the pendulum swings free and the angle of resistance after it breaks a specimen. The pendulum swings on precision ball bearings mounted on a vertical column fastened to a heavy steel base. A U-shaped hammer carries Charpy or Izod Striking Edge. The specimen holder is keyed and bolted to the base.

The energy values required to fracture a specimen subjected to impact loading are displayed on a large circular scale, color-coded for different ranges. The circular dial contains color-coded scales, one for each energy range, and illustrations on the dial show the capacity for each pendulum position.

A pointer mounted coaxially with the bearing shaft indicates the energy value. It is driven directly by a pusher arm attached to the bearing shaft that supports the pendulum. No gearing or motion transfer mechanism is involved. The SI-k3 model is designed with two capacities of 125 and 300 foot-pounds.

Precautions

- Never stand or allow anyone to stand in the path of the pendulum when it is raised.
- Never work on the impact machine with the pendulum raised. Use a block of wood to hold the pendulum away from the anvil while placement of the specimen is performed.
- Never try to latch the pendulum while it is swinging. The universal impact testing machine has a safety latch to prevent accidental latching while the pendulum is swinging.
- Always raise the pendulum by hand.

Materials

You will be provided with three types of materials:

- Aluminum or brass
- Steel
- Gray cast iron

Procedure

1. Before performing the impact test, check the zero of the pointer. Calibration adjusts the machine for friction and wind loss, so the pointer should read zero ft-lb (or zero J) after a free swing without a specimen in place:

    – Raise the safety latch and place the operating lever in the latch position. Then left the pendulum counterclockwise by hand until the latch clicks. The first click indicates the lower release position.

    – Insert the dowel pin, designed to prevent accidental application of the brake, into the hole in the head of the machine.

    – Set the pointer at maximum value for the range of your test.

    – Make sure no one is in the path of the pendulum, and then move the control lever to the RELEASE position. When the pendulum has started to swing back, remove the dowel pin and push the control lever into BRAKE position.

    – If the pointer reads zero, you are ready for your test. If not, loosen the screw that holds the pusher arm, turn the arm in the amount and direction to produce a zero reading, and re-tighten the arm.

Figure 4. Safety latch.

2. Block the pendulum away from the anvil and position your specimen. Use a block of wood to hold the pendulum away from the anvil while placement of the specimen is performed. Figure 5 shows the impact tester with the pendulum blocked.

Figure 5. Impact tester with the pendulum blocked.

3. Rest the Charpy specimen on the anvil with the notch away from the striking edge; the centerline of the notch must coincide with the centerline of the anvil.

4. Set the handle to the latch position, and then raise the pendulum until it clicks into position (click is audible). Set the pointer and pusher arm as you would to zero the machine.

Figure 6. The Charpy impact toughness testing machine—Ready position.

5. Move the handle to the release button; the pendulum would be released. Let it complete its swing through the specimen.
6. The pointer will indicate the amount of energy used to break the sample.

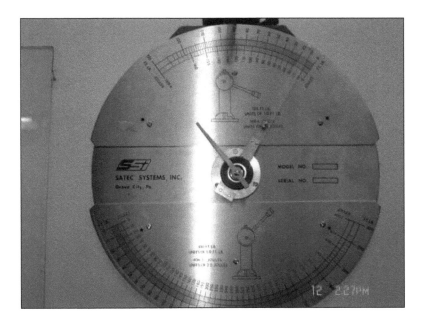

Figure 7. Reading the fracture energy.

7. Remove the pin and then rotate the handle to the break button to stop the pendulum.

8. Record the indicated energy. At this point the pointer indicates the amount of energy used to break the sample. The pusher arm, linked to the pendulum, has driven the indicating pointer to the energy value determined by the test, and then fallen back with the pendulum. Because the pusher arm and pointer work together in this way, they must be in contact at the maximum value for your test before you break your specimen.

9. Repeat this procedure for three different samples: Cast Iron, Aluminum, and Steel.

TO DO LIST

According to ASTM the laboratory report should include the following:

- The type of machine
- The type of specimen
- The temperature of the specimen
- Energy absorption of the specimen
- Type of fracture

TEST DATA:

Sample No.	Type of Specimen	Specimen Dimensions	Absorbed Energy (ft-lb/J)	Type of Fractured Surface
1	Steel			
2	Steel			
3	Steel			
4	Steel			
			Avg.=	
5	Gray Cast Iron			
6	Gray Cast Iron			
7	Gray Cast Iron			
8	Gray Cast Iron			
			Avg.=	
9	Aluminum			
10	Aluminum			
11	Aluminum			
12	Aluminum			
			Avg.=	

1. Record the exact type of steel, aluminum, and/or cast iron used in the experiment.
2. Evaluate the energy absorbed for each of the specimen in ft-lb or J. Present the results as a plot (in the form of a pie chart or any other similar plots) for all the samples at room temperature.
3. Comment on the brittle/ductile nature of the fracture surfaces.
4. What is the impact toughness for each of the three materials?
5. Correlate the impact energy with the appearance of fracture surface: brittle fracture (shiny, many small facets) or ductile fracture (fibrous, dull).
6. What is the significance/importance of the V-notch in a Charpy test sample?
7. Toughness can also be characterized by the area under a stress-strain curve for a smooth tension specimen loaded slowly to fracture. What is a significant difference in terms of the sample shape between this method and Charpy's method?
8. What else is a significant difference between this impact test and the tensile test performed in another chapter? (Hint: think in terms of loading rate.)

Note to Instructors: You can augment or edit the above procedure or To-Do List in any way that is most suitable for your lab/class situation. One of the important things about the above procedure is to ensure that your machine is calibrated. If the machine does not give a zero reading when no specimen is present, then this indicates a serious problem. Even if the machine gives a zero reading in the above case, if the striker on the hammer is not perfectly aligned with the plane of the notch on the specimen, erroneous impact energy numbers can arise.

CHAPTER 12
EFFECT OF HEAT TREATMENT ON MECHANICAL PROPERTIES (EXPERIMENT/LAB 10)

To examine the mechanical properties (hardness and impact energy, in this case) of steel due to heat treatment, samples of 4340 or A36 steel are given an austenitizing heat treatment and cooled at various rates to room temperature. The Rockwell hardness and Charpy impact energy are measured for each sample and discussed with respect to the microstructures that are expected to exist.

BACKGROUND

Review the iron–carbon phase diagram, isothermal transformation diagrams, and the heat treatment operations and properties of steel that are in your materials science textbook.

INTRODUCTION

The equilibrium phase diagram for the binary iron–carbon (Fe–C) system is shown in Figure 1. It is the basis for understanding the microstructures of the ferrous alloys (steels) and cast irons. Dissolved carbon clearly has a pronounced effect upon the liquidus, explaining why the difficulty of achieving furnace temperatures of 1600 °C caused large-scale production of cast irons to predate that of steel. Iron takes on a different unit cell structures at different temperature ranges. For example, it can be α iron (or Ferrite), which is a BCC material, at room temperature and up to 912 °C. Between 912 °C and 1394 °C it transforms to a FCC unit cell structure (austenite or γ phase). Beyond 1394 °C it goes back to BCC structure as δ-iron. α, γ, and δ are actually solid solutions of mostly iron and a small percentage of carbon atoms that dissolve interstitially in these iron allotropes. At the other end of the phase diagram is an intermediate compound Fe_3C (or cementite).

The large difference in solid solubility of carbon in austenite and ferrite, together with the existence of a eutectoid reaction (a solid phase transforming into two other different solid phases upon cooling, which happens at 0.76 wt % C in Figure 1), are responsible for the versatile behavior of steels during heat-treatment. At eutectoid composition and at a temperature above the eutectoid temperature isotherm (727 °C in the diagram), we have an austenite phase. Slow cooling enables austenite to decompose and form the micro-constituent pearlite, a lamellar composite of soft, ductile ferrite (initially at 0.022% C) and hard, brittle cementite (at 6.7% C). On the other hand, quenching of austenite from a temperature above the eutectoid temperature to room temperature results in a hard metastable phase known as martensite.

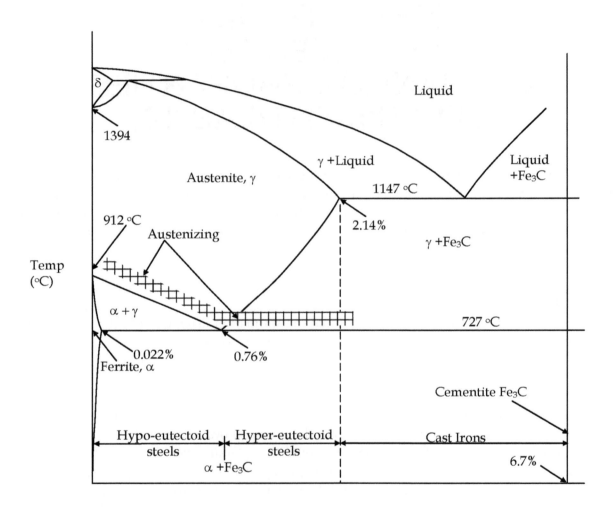

Figure 1. Equilibrium phase diagram for Fe–C system.

Eutectoid decomposition occurs in both ferrous (e.g., iron–carbon) and non-ferrous (e.g., copper–zinc, copper–aluminum) alloy systems, but it is of particular importance industrially for steels. As was mentioned above, if a Fe–C binary system was at eutectoid composition and in the austenite phase, and if this phase was cooled down slowly, a lamellar pearlitic structure of cementite and ferrite will form (alternate layers of soft and hard phases). However, when the cooling conditions are such that the alloy structure is far removed from equilibrium, an alternative transformation may occur. Thus, on very rapid cooling (by quenching to room temperature), a metastable phase called martensite, which is a supersaturated solid solution of carbon in austenite, is produced. Martensite takes the form of sharp needles in a matrix of austenite remnants. Martensite is a very brittle phase and highly undesirable for most practical reasons. Alternatively if one quenches the austenite to a temperature between 220 °C and 540 °C and then holds the temperature steady, an equilibrium phase called Bainite will develop. A bainitic structure consists of platelike grains of ferrite, somewhat like the plates of martensite, inside which carbide particles can be observed. Moreover, if one does not do a severe quench like above, but rather rapidly cools down to a temperature in the range between 540 °C and 727 °C, and then holds the material to this temperature, then the familiar pearlite structure would result. Both pearlite and bainite if held for *hours* in their temperature range (between 540 °C and 727 °C for pearlite and between 220 °C and 540 °C for bainite), they both eventually transform into a ductile microstructure called spheroidite (which has a ferrite matrix embedded within it spherical-looking cementite particles).

All of the above transformation information is contained in what is called an "isothermal transformation diagram," which is also known as a "time-temperature-transformation" plot (i.e., TTT plot). For the Fe–C system, Figure 2 shows its TTT. Figure 2 plots the time necessary at a given isotherm temperature to transform austenite of eutectoid composition to one of the three microstructures: pearlite, bainite, or martensite.

Note in this TTT that martensite formation starts at 220 °C, but only for a cooling path that is rapid and avoids the outermost nose of the C-shaped curves (the curves represent 0%, 50%, and 100% transformation to either pearlite or bainite, depending on the temperature range; see above). It is evident from such a plot that a wide variety of structures can be obtained from the austenite decomposition of a particular steel. The resulting structure may range from 100% coarse pearlite, when the steel will be soft and ductile, to fully martensitic, when the steel will be hard and brittle.

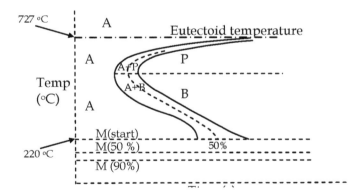

Figure 2. TTT plot for eutectoid steel. M = martensite, P = pearlite, B = bainite, A = austenite.

In most practical situations, the cooling of the austenite is continuous and the temperature is not held steady in either of the above two temperature ranges. In this case, the phases present in the TTT plot are no longer accurate transformation results. Rather, the results of transformation are martensite, pearlite, or a mixture of the two. This will depend on the exact cooling rate used. This is shown in Figure 3, which is called a "continuous cooling transformation diagram," or CCT diagram.

A cooling rate equal to the critical cooling rate or higher will produce martensite. A rate equal to 35 °C/s or slower will produce pearlite. Any rates in between these two will produce a mixture of martensite and pearlite.

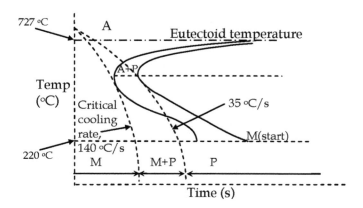

Figure 3. CCT diagram for eutectoid steel. M = martensite, P = pearlite, A = austenite.

To quench steels from hot temperature, several media or quenching materials can be used. One can use water, brine (salty water solution with 10% NaCl by weight), ice-water path, iced brine solution, conventional oils, standing air or forced air, salt, or a polymer solution. Quenching of steels increases their hardness and therefore strength. However, the quenched piece might not be spatially heterogeneous in its hardness. This is due to the exact volume or size of the piece, the quenching medium, and the characteristics of the quenching medium (e.g., still oil versus agitated oil). Agitation of the medium improves the cooling rate. Also, a smaller piece cools down faster than a bigger piece on the inside. It is very important, for engineering design purposes, to know the effect that quenching and heat-treating of materials have on their mechanical properties.

Procedure

You will be given four specimens of steel (e.g., 4340 steel). One of the steel specimens is denoted as as-received because it is in the condition that it was received from the factory/vendor. The steel samples are heated to 900 °C (to give it an austenizing heat treatment) for 60 minutes and cooled at various rates: in water (water quench); in oil (oil quench); in air (air cooled); and by leaving it in the furnace while the furnace slowly cools down (furnace cooled). Water quenching is the fastest, while furnace cooling is the slowest. Measure the Rockwell hardness of all Charpy V-notch samples provided (four indents for each specimen), and if available, examine the microstructures of polished samples in the optical microscope. Also test the impact energy of the samples.

TO-DO LIST

1. Perform an error analysis and tabulate the hardness/impact energy values measured together with the experimental error for each.
2. If microscopic observation of the specimens was not possible, describe which microstructures were most likely to have formed for each heat treatment. How do these microstructures explain the difference in

hardness following the various heat treatments of the specimens? What processing treatment was the as-received specimen most likely to have had?

3. How do your steel hardnesses compare to those of the non-heat-treated sample you tested previously? (only do for 4340.) Are your results what you expected?

4. Compare the impact energies to those you obtained from the Charpy impact test for the non-heat-treated, non-quenched sample.

5. What is a phase diagram? How do you determine what phases are present, their compositions, and the amount of each phase from a phase diagram?

6. What are TTT and CCT diagrams? How do they differ?

7. What is steel?

8. Define the terms "eutectoid," "hypoeutectoid," and "hypereutectoid." What do they refer to in steel?

9. What is cast iron?

10. What is austenite?

11. What is ferrite?

12. What is cementite?

13. What is pearlite?

14. What is martensite?

15. What is the softest structure formed by slowly cooling eutectoid steel from the austenite region?

16. What is the hardest structure formed by quenching eutectoid steel from the austenite region?

17. How does the cooling rate following an austenitizing heat treatment of steel affect its hardness?

18. How does the cooling rate following an austenitizing heat treatment of steel affect its impact energy (i.e., impact toughness)?

Note to Instructors: You can augment or edit the above procedure or To-Do List in any way that is most suitable for your lab/class situation. Exercise care when using furnaces to heat-treat material. Make sure that proper protective equipment is fully used. Also, exercise care when quenching hot metal in water or oil. Dangerous splashing can occur. Also, no one should rush to gather quenched samples quickly after dropping them in the medium. They may still be very hot. One way is to use a laser heat gun to ensure that the quenched material temperature has gone down to where it is safe to handle with bare hands.

CHAPTER 13
THE DUCTILE-TO-BRITTLE TRANSITION TEMPERATURE (EXPERIMENT/LAB 11)

This experiment will demonstrate how the impact toughness, or impact energy, is affected by temperature of the material. Specifically, the impact energy–temperature curve allows the determination of the Ductile-to-Brittle Transition Temperature (DBTT).

INTRODUCTION

Many BCC metals such as plain carbon steels undergo a change in fracture mode from ductile to brittle as the temperature is decreased. This ductile-to-brittle transition (DBT) is due to a rapid increase in yield stress and a fairly constant brittle fracture stress as the temperature decreases. Because the yield stress of FCC metals such as aluminum or many stainless steels does not increase as dramatically with decreasing temperature as BCC metals, a DBT is generally not seen in these materials.

If Charpy or Izod impact tests are conducted over a wide range of temperatures, then a plot similar to the one displayed in Figure 1 can be obtained. Such a graph illustrates the notch (impact) toughness of a material. In the figure, the impact energy of FCC metals decreases as the temperature decreases, but there is no change in fracture mode. In general, FCC materials do not have a distinct DBT.

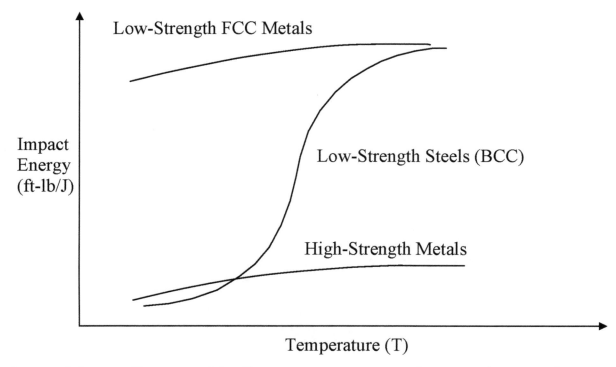

Figure 1. Schematic illustration of the Charpy impact energy curves for BCC and FCC metals.

In Figure 1, the DBTT for low-strength BCC steels is the point where the S-shaped curve changes curvature (somewhere around its middle point in this case). This changing curvature signals the transition from ductile to brittle behavior when temperature is cooled below it. Notice that with such definition, albeit graphically acceptable, this does not necessary imply a sudden change in the fractured surface nature or appearance (i.e., a sudden change in the physics of microstructural failure) once this point is crossed from right to left on the temperature axis.

Precautions

- Never stand or allow anyone to stand in the path of the pendulum when it is raised.
- Never work on the impact machine with the pendulum raised. Use a block of wood to hold the pendulum away from the anvil while placement of the specimen is performed.
- Never try to latch the pendulum while it is swinging. The universal impact-testing machine has a safety latch to prevent accidental latching while the pendulum is swinging.
- Always raise the pendulum by hand.

Materials

Two types of materials will be used:
- Aluminum (2024-T351) or brass
- Steel (any low strength steel)

Procedure

1. Use the same exact procedures from the Impact Testing chapter/experiment.
2. Redo the Charpy Impact testing for all steel and aluminum Charpy specimens at five different temperatures: liquid nitrogen temperature (77 K = -196 °C!!); ice bath temperature (0 °C); room temperature (27 °C); boiling water temperature (100 °C); hotplate heating temperature (300-400 °C).
3. BE VERY CAREFUL when handling the samples embrittled by liquid nitrogen or samples at 100 °C and 300–400 °C. Make sure you have protective heavy-duty cloth gloves on at all the times when temperature-treating the samples.
4. DO NOT rush to hold the samples after fracture by impact, as they might be very cold or very hot.

TO-DO LIST

1. Discuss the brittle appearance of fracture surface for each sample in percentage of the brittle area.
2. Determine the DBTT for both aluminum and steel.
3. Comment on brittle/ductile impact energy change as result of changing temperature for both materials.
4. Compare the fracture surfaces of samples at different temperatures with respect to the fracture surface at room temperature for both aluminum and steel.
5. According to the textbook by Callister (Introduction to Materials Science and Engineering), it is difficult to specify a single DBTT. One proposal by the textbook to identify this temperature is to associate it with a Charpy V-notch specimen impact energy of 20 J or 15 ft-lb. Using your established impact energy-temperature curve for steel, from the experimental values at five different temperatures, is this proposal correct?

Note to Instructors: You can augment or edit the above procedure or To-Do List in any way that is most suitable for your lab/class situation. Ensure that nobody touches directly, with their bare hands, the samples subjected to liquid nitrogen or boiling water/hotplate. Using the hotplate, more hot temperatures can be tested besides the two temperatures indicated above. Also, once a specimen is subjected to liquid nitrogen, it can be taken out of that environment and allowed to warm up in ambient conditions. This way, temperatures hotter than the liquid nitrogen temperature but cooler than 0 °C can be tested.

CHAPTER 14
THE NANO-
INDENTATION TEST
(EXPERIMENT/LAB 12)

This experiment will introduce and demonstrate the instrumented nano-indentation technique and the parameters that govern it.

INTRODUCTION

Indentation experiments are not strange to the engineering industry. Examining material properties using hardness tests has been used as non-destructive tests for metals for the last 100 years [1]. Indentation depends on pushing a hard indenter into the surface of the material and recording the load and indentation depth. So far we have examined two types of hardness:

- Macro-hardness/indentation: Example: Brinell, 10 mm indenter, and 3000 kg.
- Micro-hardness: In this type of testing a prescribed load is applied to an indenter in contact with a specimen, the load is then removed and the area of the residual impression is measured. The load divided by the by the area is called the hardness. Example: Vickers micro-indentation with a diamond pyramid.

In this lab, we are interested in determining nano-scale hardness. In nano-indentation test, a prescribed load is applied to an indenter in contact with a specimen. As the load is applied, the depth of penetration is measured. The area of contact at full load is determined by the depth of the impression and the known angle or radius of the indenter. The hardness is found by dividing the load by the area of contact. Shape of the unloading curve provides a measure of elastic modulus.

Researchers showed that material nano-scale hardness can be related to material stiffness (elastic modulus) and energy absorption (toughness and resilience), as well as to fracture toughness.

The recent advances in hardware control, load, and displacement measurements at the nanoscale made nano-indentation a technology that is both robust and reliable for material mechanical characterization. Nano-indentation experiments comprise loading the specimen to a specific load (usually in the range of micro- to milli-Newton (mN)), keeping the load constant for a few seconds to realize material creep (strain growth with time), and unloading the specimen leaving an indentation impression. A schematic representation of the Nanoindenter (NanoTest®) that we used for nano-indentation is shown in Figure 1.

Nano-indentation Apparatus

The NanoTest 600 is a pendulum-based nano-indentation machine developed by Micro Materials Ltd, United Kingdom. The system can be used in conjunction with the low load head which is 0.1–800 mN. The system is placed inside a thermally insulated cabinet. This cabinet serves to reduce air turbulence that would upset the pendulum, and to provide a thermally controlled environment, and also to provide sound-proofing to reduce acoustic disturbances.

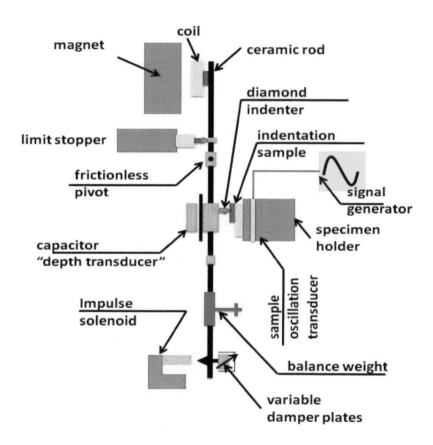

Figure 1. Schematic of the NanoTest 600 (Micro Materials Ltd.).

At the heart of the NanoTest system is a pendulum that can rotate on a frictionless pivot; see Figure 1. A coil is mounted at the top of the pendulum; with a coil current present, the coil is attracted towards a permanent magnet, producing motion of the diamond tip towards the sample and into the sample surface. The displacement of the diamond tip is measured by means of a parallel plate capacitor, one plate of which is attached to the diamond holder. When the diamond moves, the capacitance changes, and this is measured by means of a capacitance bridge. The capacitance bridge unit is located close to the measuring capacitor in order to minimize stray capacitance effects.

Sample displacement is achieved by means of three DC motors driving micrometer stages in an XYZ configuration. The motor control electronics consist of a mother board containing three power modules, an IEEE interface module, and a backlash control board. These plug into expansion slots in the computer. The motor power supply is derived by the controllers through visual interface of the system software in the computer, with motor positioning obtained by means of magnetic encoders. The motor control board communicates with the system computer via the IEEE bus. The limit stop defines the maximum outward movement of the diamond, and also the operating orientation of the pendulum, when a load is applied. Its position is manually adjusted with a micrometer. The equilibrium position of the pendulum, with zero load current, is adjusted with balance weights which are movable along both the horizontal and vertical axes.

The machine has a variety of calibrations, some of which are more often performed than others. Some of these are:

- *Load*—Establishes the forces that can be applied at the diamond tip during a measurement
- *Depth*—Relates the change in capacitance to a known distance moved by a sample in contact with the pendulum
- *Frame Compliance*—Corrects for "flexibility" in the structure of the machine
- *Microscope*—Includes a calibration for the distance from the indenter tip to the focal plane and a calibration to position the sample for precisely placed indentations

The precision of the nano-indentation apparatus is achieved by controlling and recording the time-dependent nanoscale displacement of the indenter tip as it changes with electrical capacitance. Loading is performed by sending an electrical signal to the coil, causing the pendulum to rotate about its frictionless pivot, so that the diamond indenter penetrates the sample surface. The indenter tip displacement (penetration) is measured during loading and unloading with a parallel plate capacitor that has sub-nanometer resolution. The components of the nano-indenter available in our laboratory are shown in Figure 2.

Figure 2. Components of the NanoTest System 600.

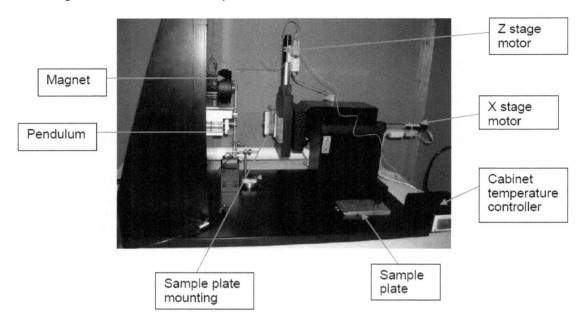

The nano-indentation tip can be of different shapes such as spherical, canonical, cube corner, and Berkovich. The Berkovich is the most popular; Figure 3 shows different indenter tips.

Figure 3. Schematic of top and side views of different indenter tips.

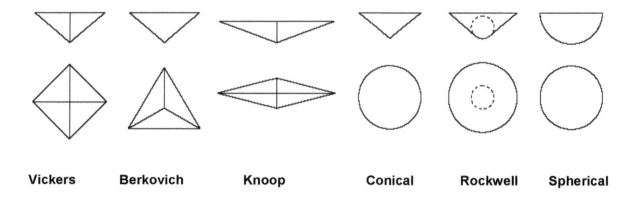

Nano-indentation Data Analysis: Projected Area

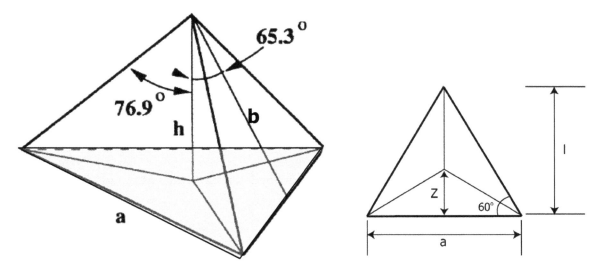

Figure 4. Geometrical parameters of a standard Berkovich indenter tip.

The standard geometry of a Berkovich tip is given in Figure 4. We will rely on the geometry prescribed in that figure to calculate an important parameter for calculating both the hardness and Young's modulus. That is the projected are $A_{proj.}$

Starting by the relation between projected length l *and the width a*

$$\tan 60^{\circ} = \frac{l}{a/2}$$

(1)

From which

$$l = \frac{\sqrt{3}}{2}a$$

(2)

Thus, the projected area can be represented as

$$A_{proj} = \frac{al}{2} = \frac{\sqrt{3}}{4}a^2$$

(3)

Utilizing the angle between the indentation depth h and the side b

$$\cos 65.27^{\circ} = \frac{h}{b}$$

(4)

Linking the geometry between the 3D figure and the 2D projection one can show that

$$h = \frac{a\cos 65.3^o}{2\sqrt{3}\sin 65.3^o} = \frac{a}{2\sqrt{3}\tan 65.3^o} \tag{5}$$

From this equation, it is easy to deduce the parameter a as

$$a = 2\sqrt{3}h\tan 65.3^o \tag{6}$$

Finally the projected are in terms of the indentation depth h can be written as

$$A_{proj} = 3\sqrt{3}h^2\tan^2 65.3^o = 24.56h^2 \tag{7}$$

In actuality one can also calculate the projected area using an electron microscopy image. For example Figure 5 shows a Berkovich indentation residue on the surface of an epoxy.

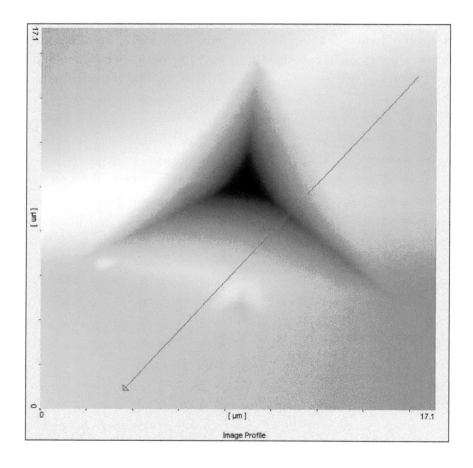

Figure 5. Sample nano-indentation projected area imaging on the surface of a structural epoxy material. The image was generated using an atomic force microscopy (AFM).

Nano-indentation Data Analysis: Hardness and Elastic Modulus

The NanoTest software uses a method developed by Oliver and Pharr for analysis [2]. This method uses the load (P) vs. displacement (h) curves to determine the hardness and elastic modulus of a given material. A typical indentation curve is shown in Figure 6. This figure also shows the values needed to do the analysis described by the Oliver and Pharr method.

There are three experimental values that are critical to find the hardness and elastic modulus. These are the peak load (P_{max}), the depth at peak load (h_{max}), and the initial unloading contact stiffness (S).

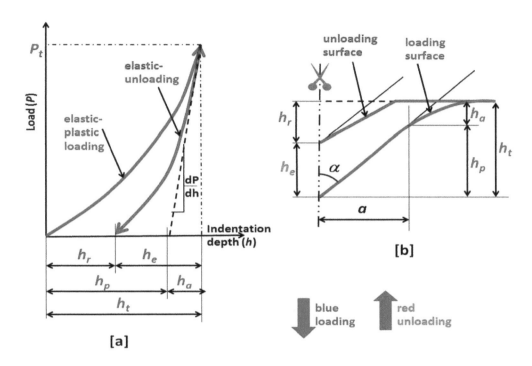

Figure 6. (a) Typical loading vs. indenter displacement curve and (b) cross-section of a Berkovich indenter trace in the sample.

The software analysis begins by fitting the unloading curve from 100% to 40% of P_{max}. This is achieved by fitting the polynomial:

$$P = \alpha(h - h_f)^m \tag{8}$$

where a, m, are constants determined by the least squares fit and h_r is the displacement when $P=0$. By differentiating this equation the contact stiffness can be derived.

$$S = \frac{dP}{dh} = \alpha m (h_{max} - h_r)^{m-1} \tag{9}$$

Once S is found the contact depth can then be calculated using:

$$h_p = h_{max} - \varepsilon \frac{P_{max}}{S} \tag{10}$$

Where a is an experimentally determined correction factor for the indenter used ($\varepsilon = 0.75$ for a Berkovich indenter tip, $\varepsilon = 1.0$ for a spherical indenter, etc. [3]). The software then uses a truncated version of Oliver and Pharr's area function

$$A(h_p) = 24.5h_p^2 + f(h_p) \tag{11}$$

Here $f(h_p)$ is a fifth-order polynomial, the coefficients of which are determined from a graph obtained from indentations done on the fused silica. This curve is called the diamond area function or DAF for short. The curve relates the area under the tip to the plastic depth. These indentations are done in a full depth range up to about one micron. For depths deeper than one micron, we assume an ideal Berkovich shape which is in good agreement with the tests done on the fused silica.

Knowing the frame compliance and the properties of the reference sample, fused silica in our study, we are able to find the projected contact area of the indenter tip at h_p. Once the stiffness and the projected area are obtained, the software calculates the effective modulus using

$$E_{eff} = \left(\frac{\sqrt{\pi}}{2} \right) \left(\frac{1}{\sqrt{A(h_p)}} \right) \left(\frac{1}{\frac{1}{S} - C_f} \right) \tag{12}$$

C_f is the frame compliance factor of the NanoTest system that is found through machine calibrations. The hardness of the sample can be computed using

$$H = \frac{P_{max}}{A(h_p)} \tag{13}$$

where A is the contact area of the indenter. So the reduced modulus can be determined from the contact stiffness of the material, which can be determined from the slope of the unloading curve, as shown in Figure 6(a).

Luckily, students do not have to carry out these lengthy calculations. Most of the nano-indentation systems are equipped with excellent data acquisition systems to measure both the load and depth continuously during loading and unloading portions of a test. These data can be then processed using software or a template embedded with the data acquisition system. For NanoTest systems, a template (made using LabView®) is available with the nano-indenter to analyze the entire load-indentation-depth curve and identify all the material properties. A snapshot of that software is shown in Figure 7.

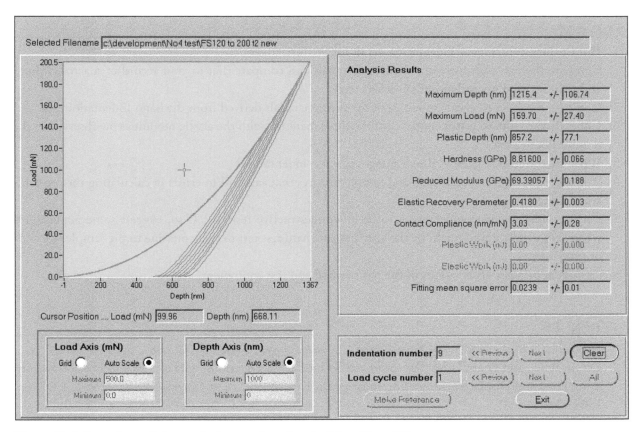

Figure 7. A Snapshot of the analysis template for the NanoTest 600.

Experiment

Here we will indent three samples using the nano-indentation: A36 steel, tool steel, and heat-treated tool steel. We will indent the three materials at 10, 50, and 100 mN loads and will find the nano-properties of the three materials using the load versus nano-indentation depth curves.

It is recommended that the samples be polished thoroughly using the procedures introduced in a previous chapter/lab.

Data Record For Nano-Indentation Depth

	A36 Steel	Tool Steel	Heat-Treated Tool Steel
Maximum depth (nm)			
Plastic depth (nm)			
Nano-Hardness (GPa)			
Reduced Elastic Modulus (GPa)			

TO DO LIST

1. Plot the load versus nano-indentation depth for the three materials.
2. Compare the nano-hardness of the three materials and compare that to your record of macro-hardness (Brinell test), and micro-hardness (Vickers test).
3. Compare the reduced elastic modulus of the three materials derived from the nano-indentation test.
4. Compare the reduced elastic modulus of the three materials with the elastic modulus you observed in the tension test of steel.
5. What are the advantages and disadvantages of nano-indentation?
6. What are some of the parameters and sample features that can lead to errors in calculating the modulus and hardness?
7. Do you consider nano-indentation to be a "nondestructive testing"? If so, suggest some applications where nano-indentation might be the only possible hardness test to carry out due to the sample size and/or properties.
8. Comment on your observations of the test method and test equipment.

REFERENCES

1. Hertz, H., "On the Contact of Elastic Solids," *J. Reine Angew. Math.*, V. 92, 1881, pp. 156-171.
2. Oliver, W.C., and Pharr, G.M. "An Improved Technique for Determining Hardness and Elastic Modulus using Load and Displacement Sensing Indentation Experiment," *J. of Materials Research*, V. 7, 1992, pp. 1564-1583.
3. Fisher-Cripps, A. C., *Nano-indentation*, Second Edition, Springer-Verlag, NY, 2004.